FIFTH EDITION

Biology
Laboratory Manual

Fred Adams, Jr., M.Ed.
Vernon L. Price, Ph.D.
Glenn E. Reynolds, M.Ed.
Trinity Valley Community College

Winston McCowan, Ed.D.
North East Texas Community College

KENDALL/HUNT PUBLISHING COMPANY
4050 Westmark Drive Dubuque, Iowa 52002

We the authors, wish to give special recognition to:

Shaine E. Morrow for the front cover photograph

David C. Pryor for the back cover photograph

Billy Meazell for drawings in this Laboratory Guide

ISBN 0-8403-9630-9

Printed in the United States of America
10 9 8 7 6 5

Mr. Rosco Francis was one of the original faculty members when Henderson County Junior College first began in the fall semester of 1946. He taught agriculture and biology and later became chairman of the Natural Science Division. He co-authored the first editions of this laboratory manual.

With utmost gratitude, admiration, and respect the authors dedicate this laboratory manual to his memory. The influence of this outstanding educator will have a lasting effect on all his co-workers and the thousands of students who were privileged to come into contact with him during his 28 years at Henderson County Junior College, now known as Trinity Valley Community College.

Contents

Preface vi
Acknowledgments vi
Introduction vii
Laboratory Instructions xi
Unit
1. The Microscope 1-1
2. Chemistry of Life 2-1
3. The Cell 3-1
4. The Algae and Bacteria 4-1
5. The Fungi 5-1
6. Liverworts, Mosses and Ferns 6-1
7. The Pine 7-1
8. Roots, Stems, and Leaves 8-1
9. The Flower 9-1
10. The Animal Cell 10-1
11. Unicellular Organisms 11-1
12. Sponge, Hydra, and Jellyfish 12-1
13. The Flatworms 13-1
14. The Roundworms 14-1
15. The Segmented Worms 15-1
16. The Crayfish and Grasshopper 16-1
17. The Starfish 17-1
18. The Clam 18-1
19. The Amphioxus and Perch 19-1
20. The Frog 20-1
21. The Rat and Pig 21-1

Preface

This laboratory manual is designed to meet the requirements of the first year of biology. It contains biological reactions, principles, and organisms from each kingdom.

This laboratory manual is easily adaptable to a general, integrated, or pure botany and zoology course of study. It is further designed to simplify the work of the laboratory which so often is a psychological barrier to the beginning student.

Great effort was taken in preparing these exercises to take advantage of specimens and materials so the operating costs can be kept to a minimum.

Acknowledgments

We, the authors, wish to give special recognition to Billy Meazell for preparing many of the drawings in this laboratory manual.

Introduction

Purpose

The purpose of this laboratory guide is to be a guide for the student to use during the laboratory period which has been set aside for the specific purpose of putting into practice what he has learned in the lecture section.

Due to the nature of the biology course, it is necessary that a student be able to use, in lab, his learned knowledge which comes from lecture. Since this is the case, it is not a textbook period, and the text is to be used strictly as a source of reference, and any other use of it will be severely penalized. This rule is to subdue the idea of using the textbook's drawings and not using the practical side of the course. In addition to this fact approximately ¾ of your grade in the laboratory section will come from a test called a "lab practical." It is necessary that a student be familiar with the slides, procedures, and techniques used in the laboratory.

For those of you who might not be familiar with the term "lab practical," we will attempt to explain what transpires during a lab test. For a period of time you will study a specific plant or animal in several aspects. For instance, you may observe an *Elodea* cell under the microscope and be responsible for labeling its parts and also learning the functions of each part (this usually comes from the lecture section). On the test, this same leaf may be set up with a cell clearly visible and the pointer on the nucleus of the cell. Such a question might be asked: What is the object the pointer is on and what is its function? With this type of question you must know more than the function of the nucleus; you must be able to identify it. This is the type of questioning involved in a "lab practical."

Materials Required to Be Furnished and Brought to Lab Each Lab Period by the Student

1. Two pencils (No.2)
2. Eraser
3. Drawing paper (unruled)
4. Notebook paper
5. Ruler
6. Textbook
7. Folder for exercises
8. Laboratory manual

ORDER OF EXERCISE

I. Fly Sheet
II. Procedure pages and questions
III. Drawings and/or plates

FLY SHEET

NAME____

EXERCISE NO.________&NAME ____________________________

DATE DUE____

BIOLOGY__________ROOM NUMBER __________

PERIOD__________

Laboratory Instructions

Value in Points

1. **Questions** 20
2. **Drawings with labels** 80
   - A. Not in correct order.
   - B. Information on fly sheet absent, incorrect order or not centered on page or incorrect information on fly sheet.
   - C. Pages missing from lab manual.
   - D. Using ink for drawing instead of pencil.
   - E. Using lined paper for drawings.
   - F. Not following instructions as given by your instructor.
   - G. Having more than 2 drawings per page **unless specified by instructor.**
   - H. Drawings in incorrect order or done in improper manner.
   - I. Not maintaining 1 inch margin around page.
   - J. Any other incorrect item not specifically covered above.

PART ONE ABOVE: If there is only 1 question on an exercise, that question will count the entire 20 points. If there are 10 questions, they count 2 points or if there are 20 questions on the exercise then they count 1 point per question.

PART TWO ABOVE: This 80 points should be divided according to the number of drawings and the number of labels involved in that exercise. If you have only 1 drawing with 1 label, that drawing and label is worth the whole 80 points. If you have several drawings with more than 1 label, the value should be divided accordingly.

ABSENCES: Students absent with unexcused absence will be docked **10 points per class day for work turned in late. No late exercises will be accepted without approval sheet.**

A grade of 100 indicates a perfect paper. No paper should receive this grade unless it is a perfect paper.

Graders will check all slides and count them at the beginning of the period and provide a sheet of paper for students to sign their names to check out slides. Count the slides in the tray, any cracked or broken slides or other damages. Graders will check to be sure that each student who checks out a slide or other equipment signs for this property.

At the end of the period, each grader will check all slides or other equipment that students have checked out making a notation of any damaged or broken slides or equipment and any slides or equipment that has not been returned. When you find damaged or broken equipment or equipment not returned, call this to the attention of your instructor immediately.

Due to variations in classification the student will complete the classification in the space provided. Refer to the text for this information.

[Use all printed capitals in your page heading.]

[Have 1 inch margin on top of page.]

[Underline page heading]

[1 inch margin]

[Do not write or draw in any margin]

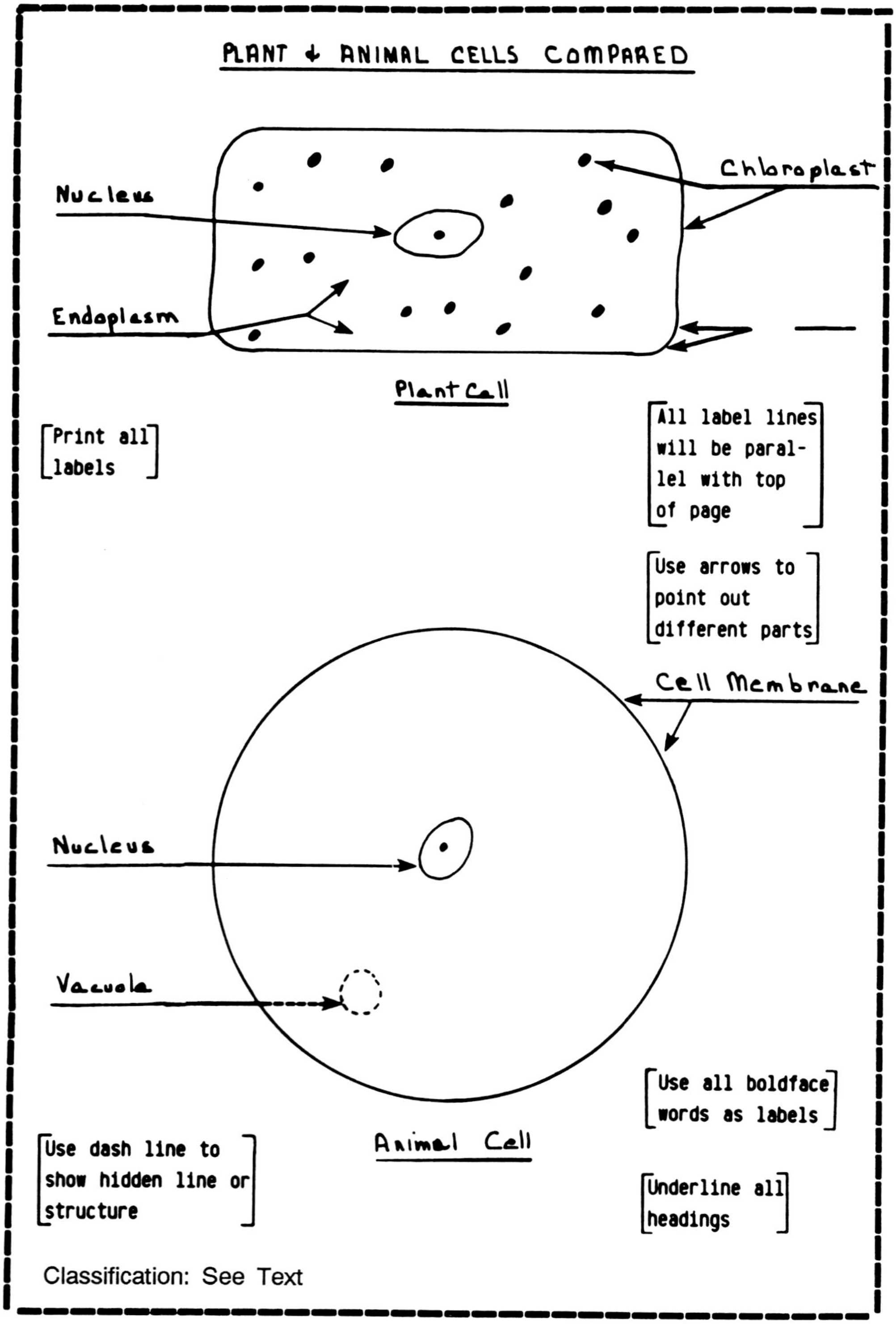

Note**Typed materials are instructions, all else examples.

[1 inch margin]

Plate 1.

Chemistry of Life

Exercise 2-A

OSMOSIS AND DIFFUSION

Introduction

Osmosis and diffusion are two prime reactions which are constantly taking place in a living organism. Osmosis is responsible for the movement of liquids from the environment to the cellular environment. Osmosis is a type of diffusion, which is usually discussed as being the movement of a gas rather than a liquid.

Materials Required

1. (HCl) Hydrochloric acid
2. Ammonium hydroxide
3. 1″ glass tubing, 14 to 18″ long
4. Absorbent cotton
5. Syrup
6. Distilled water
7. 250 ml beaker (at least one per each 2 students)
8. Semi-permeable membrane
9. Sugar
10. Potato (Irish)
11. 5% salt solution
12. 1% salt solution
13. Sufficient glassware to house the material in the exercise
14. Thistle tube or glass tubing
15. Potassium permanganate
16. Ammonium hydroxide

Procedure

1. Place piece of 1″ tubing in a horizontal position on a table. Saturate a piece of cotton with ammonium hydroxide and place it in one end of the tube. At the other end place a piece of cotton saturated with hydrochloric acid. Observe the tube for a time. Note what happened near the center of the piece of tubing.

2. Your instructor will set up an osmometer. Observe very carefully as you will have to do this also. Fill the large open end of a thistle tube with 10% sugar solution. Cover the thistle tube with a semi-permeable menbrane and submerge in distilled water. Watch the water rise in the tube.
3. Cut 3 pieces of a potato 2″ long and ¼″ square. Place each of the pieces in a separate container. To one add sufficient 5% salt solution to cover well, to another 1% salt and to yet another plain tap water. Allow to remain for several minutes. Take the pieces of potato out and examine each. Switch the pieces which were in the 5% salt and plain water and leave for several minutes. Take the pieces of potato out and examine each again. What were the results of placing the potato in the various solutions and then the results when taken from one concentration and placed in another?
4. Using a 250 ml beaker ¾ full of water, add 2 or 3 crystals of potassium permanganate and observe for several minutes.
5. Your instructor will unstopper a container of ammonium hydroxide, prepare for an odor.

Questions

1. Define osmosis and diffusion.
2. What part of a cell is the semi-permeable membrane?
3. Define homeostasis.
4. Define permeable.
5. Define impermeable.
6. Define hypotonic; hypertonic; isotonic.
7. Which procedures demonstrate diffusion? osmosis?
8. What was the role of the salt or sugar used in the procedures?
9. Define capillary action.
10. Define cohesion and adhesion.

Exercise 2-B

SEPARATION OF LEAF PIGMENTS

Introduction

In the process of life the plant is constantly producing carbohydrates. One of the prime factors in producing these carbohydrates is the availability of chlorophyll.

The prime objective of this experiment is to see the other pigments which are in the leaves of a plant, though oftentimes not seen due to the large amounts of chlorophyll.

Materials Required*

1. Developing Chamber
2. Chromatography Strips
3. Developing Solvent
4. Chlorophyll and Carotenoid Extract
5. Capillary Tubes

*All materials are available in commercial kits.

Procedure

1. Place 20–25 drops of Solvent in the **developing chamber** and replace cap.
2. Place a Chromatography Strip on the table. Using a capillary tube pick up Plant Pigment from the container. (Be sure to keep your finger on the end of the tube to control the size of drop) Place a series of drops across the white surface of the strip about 1/2 inch from the bottom. Allow to dry and repeat 3 times allowing each to dry before the next layer is added.
3. Repeat procedure 2 using the pigments from the second container if both Chlorophyll and Carotenoid extracts are available.
4. Carefully place the dry strip in the developing chamber and observe. Do not disturb. The **Solvent** will rise through the strip carrying pigments with it. Each pigment will stop at a different level on the strip.
5. Remove the strip from the chamber when the Solvent nears the top of the strip. (Approximately 1 inch) Mark upper progress of solvent with a pencil IMMEDIATELY upon removal from Developing Chamber.
6. Draw **Chromatography strip** showing **start point,** stop point for each color, and the solvent point marked in procedure 5. Draw and label each **Color** at the proper place on the drawing and identify with **Pigment Name** and **Pigment Travel** (in mm) and **Rf.** (see chart 1)
7. Identification of pigments may be determined by the Rate of Flow (Rf) and Pigment Name. (see chart 1) Rf is determined by the following formula:

$$Rf = \frac{\text{Pigment travel}}{\text{Solvent travel}}$$

   A. Measure the distance the solvent traveled on the strip between the start point and the mark made in procedure 5.
   B. Measure each pigment from start point to stop point.
   C. Divide the Pigment travel by the Solvent travel.

8. Use the color and the Rf number to identify the pigments from chart 1. Record both the Rf number and the name of the pigment by each of the colors on the drawing done in procedure 6.

Chart 1

Color	Pigment Name	Distance Traveled	Rf Number
Yellow	Carotene		.98
Yellow orange	Alpha Carotene		.97
Yellow orange	Beta Carotene		.94
Yellow	Xanthrophyll		.86
Red/pink	Lycopene		.81
Red/pink	Xanthrophyll		.80
Yellow	Leutein		.75
Dark Grey	Phaeophytin a		.67
Yellow	Violaxathin		.66
Light Grey	Phaeophytin b		.60
Yellow	Xanthophyll		.53
Yellow	Xanthrophyll		.50
Lt. Blue-Green	Chlorophyll a[1]		.48
Dk. Blue-Green	Chlorophyll a		.46
Lt. Yellow-Green	Chlorophyll b[1]		.30
Yellow	Neoxathin		.28
Dk. Yellow-Green	Chlorophyll b		.25
Yellow	Xanthrophyll		.15

Your Rf numbers will vary from these. Use the calculations from your drawing.

Questions

1. Define Solvent
2. Define Solute
3. Define Filtrate
4. Define Precipitant
5. How many different colors are seen in a leaf during the spring or summer?
6. Why are the other colors not seen?
7. What causes the loss of the green color and the appearance of the other colors in the fall?
8. On your chromatography strip which is the darkest?
9. On your chromatography strip how many bands were chlorophyll?
10. On your chromatography strip how many bands were present?

Exercise 2-C

CARBOHYDRATES

Introduction

Carbohydrates are molecules composed of carbon, hydrogen, and oxygen in a 1:2:1 ratio. They can be classified as monosaccharides, disaccharides, or polysaccharides. Carbohydrates serve as a readily available fuel for most organisms.

Purpose: To determine the types of carbohydrates that are found in various foods.

Two simple chemical tests can be used to detect the presence of carbohydrates:

1. The presence of certain carbohydrates called "reducing sugars" can be detected with Benedict's solution.
2. The presence of starch and glycogen can be detected with Iodine.

Materials

1. Solutions: dextrose, lactose, sucrose, starch and corn syrup.
2. Extracts: celery, carrot, potato and pear.
3. Test tubes
4. Glass plates
5. Benedict's solution
6. Iodine

Procedure

1. Reducing sugar test: Place about 2 ml of each solution or extract in a separate test tube. Place 2 ml distilled water in a tube for use as a control. To each tube add 5 drops of Benedict's solution. Heat the mixture in a hot water bath for 2–3 min. Observe any color changes.
2. Starch test: Place 3 drops of each solution or extract on a drop plate. Add 1 drop of Iodine solution to each sample. Observe any color changes.
3. Results: Record your results on a chart. Identify the type of test, simple, observations, interpretations.

Exercise 2-D

FATS

Introduction

A fat is one of the fundamental biological molecules produced from a reaction between glycerol and 3 fatty acids. The fat is a hydrophobic (water fearing) molecule. The type of hydrocarbon group remaining on the fatty acid can be detected using the dye Sudan III.
Purpose: To detect the presence of fats in some foods.

Materials

1. Sudan III
2. Oil
3. Test tube

Procedure

1. Brown paper test: One test for the presence of fats is to simply place a drop or two of the sample to be tested on a piece of brown paper. Allow the paper to dry. The presence of fat is indicated by a translucent spot on the paper. You may need to hold the paper to a light source to observe.
2. Sudan III test: Sudan III is a stain that tests for the presence of fat. Add 2 or 3 ml of the food to be tested to a test tube. Add 2 ml of water and 3 drops of Sudan III. Shake the tube gently. The presence of fat will be indicated by a red ring as the lipids rise to the top and are suspended in globules and are stained.

Exercise 2-E

PROTEINS

Introduction

Proteins are complex polymers of amino acids. They are composed of carbon, hydrogen, oxygen, nitrogen, and sometimes sulfur.
Purpose: To detect the presence of proteins in some foods.

Materials

Egg albumin, gelatin, milk, chicken broth, distilled water, 0.02 M Copper Sulfate ($CuSO_4$), 6 M Sodium Hydroxide (NaOH), concentrated Nitric Acid (HNO_3), test tubes.

Procedure

1. Biuret test: Label the test tubes with solutions to be tested. To 2 ml of each solution add 2 ml of 6 M NaOH and 4 drops of 0.02 M $CuSO_4$ and shake the tubes to mix the solutions. The appearance of a violet pink color indicates the presence of proteins and specifically the presence of the peptide bond between the amino acids.
2. Xanthoproteic test: Label the test tube with the names of the solutions to be tested. To 1 ml of each solution, add 5 to 10 drops of concentrated HNO_3. BE VERY CAREFUL—NITRIC ACID IS VERY CORROSIVE. The appearance of a yellow color indicates the presence of proteins and specifically the benzene nucleus.
3. Results—Record your results on a table.

Exercise 2-F

ENZYMES

Introduction

Enzymes are organic catalysts found in living cells. To date, all enzymes that have been isolated have been found to be protein. Enzymes are important in catalyzing all types of metabolic processes in cells.

Purpose: To compare the actions of an enzyme known as catalase to a nonprotein catalyst under different conditions.

Materials

3% hydrogen peroxide, manganese dioxide, liver, potato, sand, test tubes.

Procedure

1. Catalytic reactions—Add 2 ml of hydrogen peroxide to two test tubes. Place 0.01 g sand in one test tube and 0.1 g of manganese dioxide to the second tube. Observe the results and record the rates of reaction.
2. The effect of an enzyme—Add 2 ml of hydrogen peroxide to each of two clean test tubes. In one place a small piece of liver and in the other a small piece of potato. Record the rates of reaction and compare these results to those with the manganese dioxide. DO NOT DISCARD THESE MATERIALS.
3. Reusing an enzyme—Divide the liquid portion of the previous tube in #2 containing the liver into two test tubes. Cut the liver from #2 into two equal portions and add these to the two test tubes. To the first test tube add a fresh piece of liver and to the second test tube add 1 ml of hydrogen peroxide. Record your observations and explain the reaction in the test tube containing the fresh liver. What do you think would happen if additional hydrogen peroxide were added to the second test tube?
4. Effect of the size of particles—Place a small piece of liver in one test tube and a small piece of potato in a second. Add a pinch of sand to each tube and crush these materials with separate stirring rods. Add 2 ml of hydrogen peroxide to each tube and then observe and record the rates of these reactions. Compare the results with those in #2.
5. Results—Record your results in a table. Designate the rate of the reactions as: no reaction, slow, moderate, fast, very fast.

Questions

1. How do you account for the differences in the rates?
2. Can hydrogen peroxide be broken down by catalysts other than those found in living systems? Explain.
3. Describe the effect of particle size on the rate of enzyme reaction.
4. Are enzymes reusable?

UNIT 3

The Cell

Exercise 3-A

CELL TYPES

Introduction

The structural units of plants are known as cells. In this exercise we will study several types. Note the differences and the similarities in each type.

In most plant cells you will be able to find the **cell wall, cytoplasm, nucleus,** and **vacuoles.** However, structural differences may occur in different cell types.

Materals Required

1. *Elodea* leaf
2. 5% salt solution
3. Knife
4. Iodine
5. Onion
6. Pear
7. Microscope

Procedure

1. Obtain a leaf from the terminal end of an *Elodea* plant.
   A. Mount the leaf in a drop of water on a clean slide and cover.
   B. Examine under low power of your microscope.
   C. Now examine under high power and focus on the leaf where the individual cells of only one layer may be studied in detail. Draw one cell using high power objective.
   D. Observe the movement of the **cytoplasm** in your cell.
   E. Note the heavy **cell wall** around the cell.
   F. The **plasma membrane** is located just below the cell wall, but you will not be able to distinguish it from the wall. The plasma membrane is the outer part of the **cytoplasm**.
   G. The center of the cell will be occupied by a large **vacuole**.
   H. Within the cytoplasm you will be able to see green, spherical **chloroplasts**.
   I. The **nucleus** of the cell is found in the cytoplasm between the cell wall and the vacuole. The nucleus is so near the color of the cytoplasm it is sometimes necessary to stain the cell to find it.

1. Staining can be done by using iodine. Place a drop of iodine at the edge of the cover glass and touch a blotter to the opposite edge of the cover glass.

2. Obtain another leaf and mount it in a 5% salt solution and study under the microscope.

   A. After a time water will be removed from the cell by exosmosis.

   B. As the water is being removed, the protoplasm shrinks. As the protoplasm shrinks, it pulls the **plasma membrane** away from the cell wall. The plasma membrane now appears as a membrane around a ball of protoplasm in the center of the cell. This ball includes all parts of the cell except the **cell wall.**

      1. The action of salt on the cell's parts is called plasmolysis and the cells are said to be plasmolyzed.

   C. Draw a **plasmolyzed cell** and label.

3. Next peel off the thin outer cell layer, the epidermis, from the surface of the *Allium* (onion). Take a small piece of the epidermis and prepare a water mount, be careful not to fold the epidermis.

   A. Observe the general appearance of the cells under low power and then change to high power and study details.

   B. It will be necessary to stain the cell with iodine as was done to the *Elodea* cell.

   C. Draw one cell of the onion and label all parts that are found in both the *Elodea* and onion.

4. Obtain a section cut from a pear.

   A. Scrape cells from the cut surface and prepare a water mount.

   B. Observe under high power. Note the spherical to oval, thick-walled **Stone cells**. These cells are distributed in groups among the living cells in the pear.

   C. The Stone cells have a secondary cell wall and give the pear its gritty texture.

   D. Draw a cluster of stone cells under high power and label the thick, heavy **cell walls.**

Questions

1. What is the function of the plasma membrane?
2. What is the function of the cell wall?
3. Identify three types of plastids and give their function.
4. Why can the edible part of an onion contain chlorophyll?
5. What is the difference in the chemical make up of the cell wall and plasma membrane?

Exercise 3-B

PLANT CELL MITOSIS

Introduction

All living plants have the power to grow. In living plants this is accomplished by the addition of new cells. This addition of cells is the result of two processes: The first is a nuclear division known as mitosis and the second is a cytoplasmic division known as cytokinesis. Mitosis results in the formation of two identical nuclei and cytokinesis occurs when the new cell wall is formed after mitosis is completed.

This is a continuous process but for study it has been broken into six phases.

Materials Required

1. Prepared mitosis slides
2. Microscope

Procedure

Obtain a prepared mitosis slide and focus under low power.
Locate all phases under high power.
Make detailed study and drawings under high power.

1. Interphase: Locate a cell containing a nucleus bounded by a nuclear membrane in which there is a chromatin network (no actual chromosomes visible).
   A. This cell is in Interphase which is a stage between divisions.
   B. Draw and label. Include the following parts in labeling: **cell wall, cytoplasm, nucleus, nuclear membrane,** and **chromatin network.**
2. Prophase: Locate a cell in which the chromatin network has formed rod-like chromosomes. The chromosomes should be scattered and there will be no nuclear membrane. A spindle figure is also present.
   A. This cell is in the first phase of nuclear division known as Prophase.
   B. Draw and label. Include the following parts in labeling: **cell wall, chromosomes,** and **cytoplasm.**
3. Metaphase: Locate a cell in which the chromosomes have lined across the center of the cell.
   A. This cell is in Metaphase.
   B. Draw and label. Include the following parts in labeling: **cell wall, chromosomes, spindle figure,** and **cytoplasm**.
4. Anaphase: Locate a cell in which the chromosomes have separated into two groups.
   A. This cell is in Anaphase.
   B. Draw and label. Include the following parts in labeling: **cell wall, chromosomes, spindle figure,** and **cytoplasm**.
5. Telophase: Locate a cell in which the chromosomes have formed two tight groups one at each end of the cell.
   A. This cell is in Telophase.
   B. Draw and label. Include the following parts in labeling: **cell wall, chromosomes, cell plate** and **cytoplasm**.

6. Daughter cells: These are the two cells that have been formed by completion of the cell wall (cell plate). They will be only 1/2 as large as the other cells.
   A. Draw the Daughter cells.
   B. Label the **cell wall, cytoplasm, nucleus, nuclear membrane** and **chromatin network** of each **daughter cell**.

Questions

1. When are the chromosomes duplicated in mitosis?
2. What is the role of the spindle fibers?
3. When does the nuclear membrane disappear? reappear?
4. What is the relationship between all of these terms: DNA; chromatin network, gene, chromosome?
5. What is the function of DNA in a cell?

Exercise 3-C

MEIOSIS

Introduction

In the life of plant or animal, in order for the species to be continued, there must be reproduction. In many organisms there is only asexual reproduction, which does not involve the fertilization of an egg. On the other hand, quite a few plants and animals reproduce in a sexual manner in which an egg is fertilized by the male sperm. Sexual reproduction brings about the improved appearance of many organisms as well as giving hybrid vigor.

In the development of both egg and sperm a cellular process takes place whereby the chromosome number is reduced by ½ until the union of sperm and egg, which replaces the original number of chromosomes. In the case of the human the original number of chromosomes is 46. When the sperm and egg develop each contains only 23 chromosomes until fertilization.

Materials Required

1. Prepared slide of phase I and phase II of meiosis
2. Microscope

Procedure

1. By use of the prepared slide you should become familiar with the appearance of the cell as it changes from one stage to another.
2. Prepare a drawing of individual cells in each of the following stages, labeling the following parts: **chromosomes, spindle figure, cell plate.**
   - A. **Interphase I**—Will appear as a single cell with the same characteristics as in Interphase of mitosis.
   - B. **Prophase I**—Again will have the same characteristics as Prophase of mitosis.
   - C. **Metaphase I**—Will also have the same characteristics as Metaphase of mitosis.
   - D. **Anaphase I**—The **spindle figure** will be very pronounced, but still having Anaphase characteristics of mitosis.
   - E. **Telophase I**—Ends the first phase of meiosis, having very similar physical appearances as Telophase of mitosis.
   - F. **Prophase II**—There is no daughter cell development in meiosis and Prophase II will appear as 2 cells going through cell division while enclosed in a single cell wall (or membrane). This is characteristic of all the second phase of meiosis, all taking place in a single enclosure (in some cases two enclosures).
   - G. **Metaphase II**—Will be 2 distinct sets of chromosomes lining up along the equatorial plate (cell plate). In the second phase of meiosis the angle to the original happenings.
   - H. **Anaphase II**—Will produce 4 sets of chromosomes migrating toward the individual poles. Still within one enclosure.
   - I. **Telophase II**—Completes the phase of meiosis and will appear as a single cell with 4 nuclei inside.

Questions

1. What is karyokinesis?
2. Explain the difference between a chromatid and a chromosome.
3. How does adenine, cytosine, guanine, and thymine function in meiosis?
4. Who are Watson and Crick and what was their theory concerning chromosome duplication?

The Algae and Bacteria

Exercise 4-A

THE BLUE-GREEN ALGAE

Kingdom ________________ Phylum/Div. ______________________ Class ______________

Introduction

The blue-green algae includes a large group of simple green organisms. The protoplasm has very little structural differentiation. The cells of the blue-green algae do not contain organized nuclei, but granules of nuclear material are present. Chlorophyll is found diffused in the outer parts of the protoplasm and plastids are not present. In addition to green chlorophyll a bluish pigment, phycocyanin, is found which gives these plants their blue-green color.

The members of this group show a wide variety of structure. Some members are unicellular. In others the cells may form colonies that may be filamentous, ball shape or other forms. Mucilaginous sheaths are often found around these organisms.

In this exercise you will study one representative of the unicellular form, the *Gleocapsa,* one filamentous form, the *Oscillatoria,* and one colony in which the cells form chains inside a mucilaginous ball, the *Nostoc.*

Materials Required

1. Living or preserved specimens
2. Prepared slides
3. Slides and cover glasses
4. Microscope

Procedure

1. Obtain a prepared slide of *Gleocapsa* and observe under the microscope.
   A. Find and study individual cells.
   B. Observe the mucilaginous sheath, it will appear as a clear area around the cell wall.
   C. Note the shape and size of the cell.
   D. Under high power study the structure of the cell. Note the absence of the nucleus and chloroplasts.
   E. Draw several cells and label the **cell wall, mucilaginous sheath,** and the **protoplasm.**
   F. Observe cells which are elongated and constricted in the middle, these cells are undergoing fission.
   G. Draw and label cells in various stages of fission using high power. Use the labels in part E.

2. Obtain a prepared slide of *Oscillatoria* (if available use living or preserved specimens).
   A. Study under low power of your microscope.
   B. Note the general shape of the filaments.
   C. Study the size, shape and arrangement of the cells.
   D. Study the filament under high power.
   E. Note the absence of nuclei and chloroplasts.
   F. Carefully study the cells at the ends of the filament. Note any differences from the other cells of the same filament.
   G. Draw and label a filament. Include the following labels: **cell wall, protoplasm, filament** and **cell.**
3. Examine a specimen of *Nostoc,* note the appearance and texture of the gelatinous balls.
   A. Obtain a specimen and study under the microscope.
   B. Note the shape, size and arrangement of the cells of the colony.
   C. Under high power note the absence of nuclei and chloroplasts.
   D. Carefully study the colonies and look for slightly enlarged, colorless cells, known as heterocysts, scattered irregularly along the filament.
   E. Draw and label a colony of *Nostoc,* include at least one **heterocyst, mucilaginous sheath, cell wall, protoplasm** and **colony of cells.**

Questions

1. What is the name of the bluish pigment?
2. What are the general characteristics of the Kingdom Monera?
3. Are gametes produced by the Blue-Green Algae?
4. Explain how fission and fragmentation relate to *Oscillatoria* and *Nostoc.*
5. Define: Unicellular and Colonial.

Exercise 4-B

THE BACTERIA

Kingdom ________________ Phylum/Div. ______________________ Class ______________

Introduction

The bacterium cell is found almost everywhere. Some have even been found by scientists in space exploration where no life was thought to exist.

The bacteria have a wide range of sizes and shapes. The 3 basic shapes are: cocci—round, bacillus—rod shaped, and spirillum—spiral or cork screw shape.

Material Required

1. Petri dishes with sterile agar
2. Wax marking pencils

Procedure

1. During one class period your instructor will bring sterile petri dishes with nutrient agar and allow you to innoculate them. Divide the bottom plate (one with agar) into 4 equal parts. Number each section and record what is placed on each. Identify top plate (lid) with table number and class period. After this is done the dishes will be placed, inverted, in the incubator for 24 to 36 hours at 37°C.
2. The next class period there should be many bacterial colonies present. Draw petri dish as observed showing all growth. Label the colonies as to their source. Refer to your record of innoculation.

Questions

1. Define: Antiseptic, disinfectant, bactericidal, strep, and staph as they relate to bacteria.
2. What were the contributions of Koch and Lister?
3. What is the role of a bacterial spore?
4. Define: pathogen, germ.
5. Define Brownian movement as related to bacteria.

Exercise 4-C

THE GREEN ALGAE

Kingdom ________________ Phylum/Div. ______________________ Class ____________

Introduction

Species of *Chlorophyta* are most numerous in fresh water and all possess distinct nuclei, chloroplasts, and cell walls. Some species are unicellular. In others colonies are formed, in still others the bodies are multicellular. Most green algae are so small that they can be seen only with a microscope; a few are large enough to be readily seen by the naked eye.

Materials Required

1. Microscope
2. Piece of bark or wood showing growth of *Protococcus (Pleurococcus)*
3. Slide showing *Protococcus (Pleurococcus)*
4. Slide showing *Chlamydomonas*
5. Living or preserved filaments of *Spirogyra*
6. Slide with vegetative branches of *Spirogyra*
7. Slide with *Spirogyra* in conjugation
8. Slide of *Oedogonium*
9. Slide of *Volvox*
10. Living or preserved *Volvox*

Procedure

1. *Protococcus (Pleurococcus)*
   - A. Obtain a piece of bark or wood that contains some living *Protococcus* on it. Scrape off a little of this green material and prepare a water mount on a clean glass slide.
   - B. Observe under the microscope, first under low power, then high power.
   - C. Notice the shape of the cell and try to identify the **cell wall, chloroplast,** and **nucleus.**.
   - D. Draw and label several groups of *Protococcus,* showing the different ways the single cells group together.
   - E. Obtain a prepared slide of *Protococcus (Pleurococcus).*
   - F. Note the parts of the plant previously mentioned.
   - G. Draw and label the *Protococcus* as viewed on the prepared slide.
2. *Chlamydomonas*
   - A. Obtain a prepared slide of *Chlamydomonas.*
   - B. Try to identify the following parts: **flagella, chloroplasts, nucleus, pyrenoid, eye spot,** and **cell wall.** (All visible may be **nucleus** and cell wall).
   - C. Draw and label *Chlamydomonas* showing the structures indicated.

3. *Spirogyra*
   A. Prepare a water mount of the living or preserved material of *Spirogyra.* Study these under low power. Under high power locate and identify the following: **spiral chloroplasts, pyrenoids** (imbedded in chloroplasts), **nucleus,** strand-like **cytoplasm**, and the **cell wall.**
   B. Draw and label a branch of *Spirogyra.*
   C. Obtain a prepared slide of *Spirogyra* showing vegetative branches.
   D. Compare this with living material previously studied.
   E. Obtain a prepared slide of *Spirogyra* showing conjugation.
   F. Under low power locate and draw the following:

      Protuberances of the cell walls of adjacent filaments in which the cell contents have contracted. This is the formation of a **conjugation tube.** The moving of **gametes** from one filament through the conjugation tubes to unite with the protoplasts in the other filament to form **zygotes. Zygospores**—the zygotes secrete thick walls.
4. *Oedogonium*
   A. Obtain a prepared slide (or make water mount depending on which is available).
   B. Locate and identify the **nucleus, chloroplast,** and **pyrenoids.**
   C. Draw and label a portion of the branch of *Oedogonium.*
   D. On the prepared slide or living material locate the sex organs; the **antheridium** and **oogonium.**
   E. Draw and label a section of plant body showing only the sex organs.
5. *Volvox*
   A. Obtain a prepared slide of *Volvox,* a motile colonial form of green algae.
   B. Under low power, you will observe that each colony is a hollow sphere.
   C. Located inside of the spheres are the daughter colonies. Certain cells of *Volvox* will increase tremendously in size, and by a series of longitudinal divisions form daughter colonies.
   D. Draw a *Volvox* colony and label **colony of cells** and **daughter colony.**

Questions

1. Explain the ecological role of the zygospore in the life cycle of *Spirogyra.*
2. Describe the habitat conditions of *Protococcus.*
3. Define: Antheridium and Oogonium.
4. Define: Conjugation.
5. What is the function of the pyrenoids?

Exercise 4-D

GOLDEN-YELLOW ALGAE

Kingdom ________________ Phylum/Div. ________________________ Class ______________

Introduction

This group includes chiefly unicellular plants. The most important members are known as *Diatoms*. The *Diatoms* are found in almost any body of water the world over, both fresh and salt. The *Diatoms* have walls made of silicon that bear characteristic markings. These fine lines are arranged in geometrical patterns. Another example of the golden-yellow algae is a filamentous form, lacking dividing cross-walls, called *Vaucheria*.

Materials Required

1. Prepared slides of *Diatoms*
2. Living *Diatoms* (optional)
3. Microscope
4. Prepared slides of *Vaucheria*

Procedure

1. Obtain a prepared slide or make a water mount of living *Diatoms*.
   A. Study the slide under the microscope finding as many different types as possible.
   B. Draw at least 8 different types of *Diatoms*.
2. *Vaucheria*
   A. Examine a water mount of living *Vaucheria* (if available). Notice that the plant body of *Vaucheria* has no transverse walls except where the reproductive organs are cut off from the rest of the thallus.
   B. Draw and label **cell wall, nuclei,** and **chloroplasts** in a **coenocytic filament.**
   C. Obtain a prepared slide of *Vaucheria* and examine under high power.
   D. Find the large pair of **oogonia** each with an **egg**.
   E. Find the elongated **antheridium** with many **sperm.**
   F. Draw and label the sex structures of *Vaucheria*.

Questions

1. List 6 commercial uses of *Diatoms*.
2. Describe diatomaceous earth and name the principal area where it is mined.
3. Why are different sizes of the same shape *Diatom* viewed?
4. Describe the role of auxospores in the life cycle of *Diatoms*.
5. What is meant by the term coenocytic?

Exercise 4-E

BROWN ALGAE

Kingdom ____________________ Phylum/Div. ____________________________ Class _______________

Introduction

The brown algae receives its name from the presence of an abundance of a brown pigment. This phylum includes some of the largest and most complex algae. This group is made up chiefly of multicellular marine algae. In this exercise you will study the brown algae, *Fucus*.

Materials Required

1. Preserved specimen of *Fucus*
2. Prepared slides of the receptacles
3. Microscope
4. Textbook

Procedure

1. Obtain a preserved specimen of *Fucus*.
   A. Note the flat, branched body or **thallus.**
   B. The stem-like part of the thallus is known as the **stipe**.
   C. At the lower end of the stipe locate the **holdfast.**
   D. The flattened upper leaf-like structures are known as **blades** and bear gas-filled enlargements known as **bladders.**
   E. Next study the inflated tips of the blades, these are the **receptacles.** The receptacles are covered with small, blister-like pores that lead into cavities known as **conceptacles.** The **gametangia** are produced inside the **conceptacles.**
   F. Note the Y-shaped branching of the thallus, this type branching is known as **dichotomous branching.**
   G. Draw and label the *Fucus*.
2. Obtain a prepared slide of the **female conceptacle.**
   A. Locate the **oogonia** with eggs on a short **stalk.**
   B. Locate the **sterile hairs** that separate the oogonia.
   C. Draw and label the cross section of the female conceptacle, including several oogonia and the sterile hairs.
3. Obtain a prepared slide of the **male conceptacle.**
   A. Locate the branches bearing the **antheridia** with **sperm.**
   B. Draw and label the male conceptacle.

Questions

1. What is the function of the: holdfast, air bladders, receptacles?
2. Describe the fertilization process of the brown algae.
3. Explain alternation of generation as seen in some brown algae.
4. Identify 3 commercial uses of the brown algae.
5. What is the difference between the receptacles and conceptacles in the *Fucus*?

UNIT 5

The Fungi

Exercise 5-A

BREAD MOLD

Kingdom ________________ Phylum/Div. ______________________ Class _____________

Introduction

The bread mold will be studied as a representative of all the molds although it does not possess all the characteristics associated with the molds.

Rhizopus is a common bread mold which grows rapidly on breads without preservatives and will be used for this study.

Materials Required

1. Bread which does not have preservatives added
2. Soil to serve as innoculant
3. Incubator to maintain proper temperature
4. Slides

Procedure

1. Approximately 2 days prior to the time of use, bread should be innoculated with a minute amount of soil in order to begin the growth of the mold.
2. Observe the **vegetative hyphae** noting the **coenocytic condition** (lacking cross walls and are multinucleate), the short branched hyphae penetrating the substratum called the **rhizoids,** the upright hyphae or **sporangiophores** with expanded circular ends the **sporangia,** and the hyphae parallel to the substratum connecting the colonies the **stolons.**
3. Locate and draw the sexual reproduction (conjugation). **Sexual branches, gametes, zygote, zygospore.**

Questions

1. Define: Mycelium
2. Name three different hyphae found in bread mold.
3. What type or types of reproduction are present in bread mold?
4. Name a product produced by mold.
5. What is the blue green mold found growing on bread, citrus fruits and other organic matter?

Exercise 5-B

YEAST

Kingdom ________________ Phylum/Div. ______________________ Class ____________

Introduction

Yeast are fungi but the body is not composed of hyphae. The entire plant consists of a single spherical or oval cell enclosed by a thin cell wall.

For this study common baker's yeast will be used.

Materials Required

1. Yeast cake
2. Sugar or syrup solution
3. Fermentation tube
4. Slides and cover glasses
5. Microscope

Procedure

1. At the first of the period yeast will be added to a 10% sugar or syrup solution in a fermentation tube.
   A. Observe this tube and note the gas rising and being trapped in the closed end of the tube.
   B. Note that the bubbles are produced in the absence of oxygen, in the closed end of the tube, as well as at the open end of the tube where oxygen is present.
   C. Make a drawing of the **fermentation tube** showing the top and bottom **level before fermentation** of the **10% sugar-water-yeast solution.** After fermentation label the location of **alcohol, carbon dioxide** and the top and bottom **level after fermentation.** Indicate with arrows the direction of liquid movement.
2. Obtain a sample of yeast solution from your instructor.
   A. Study this material under the microscope.
   B. Note the large vacuole occupying most of the space in the center of the cell.
   C. Draw a yeast cell. Label the **cell wall, vacuole** and **protoplasm.**
   D. Next find the cells that are in the process of budding. These will appear to be two cells of different sizes connected to each other.
   E. Draw and label three cells budding. It is not necessary to include all internal structures on each cell. Label the **bud, cell wall, vacuole** and **protoplasm.**

Questions

1. What are the end products of fermentation?
2. List two commercial products of yeast.
3. What is the principal method of reproduction in yeast?
4. Are all yeasts beneficial? Explain.
5. Define fermentation.

Exercise 5-C

LICHENS

Fungi-Kingdom ____________ Phylum/Div. ____________ Class ____________

Algae-Kingdom ____________ Phylum/Div. ____________ Class ____________

Introduction

Lichens are not single organisms, but a mutual benefit association between two plants.

Materials Required

1. Lichen specimen—3 types
2. Prepared slides

Procedure

1. Lichen bodies or thalli are of three types: Crustose, in which the thallus lies flat; Foliose, in which the thallus has leaf-like lobes; Fruticose in which the thallus is branched with upright branches. Make a drawing of each of the three types identified above.
   A. Obtain several specimens and classify as to thallus type.
   B. Be sure you have studied at least one specimen of all types available.
2. Obtain a prepared slide of the cross-section of a lichen thallus.
   A. Make a drawing of the **thallus** outline and label **fungi** and **algae (protococcus).**

Questions

1. Explain the symbiotic relationship of the two plants that make up lichens.
2. What is the method of reproduction used by the fungi that make up a lichen?
3. What is the method of reproduction used by the algae that make up a lichen?
4. Define mutualism.
5. Explain how lichens can live and grow in habitats and situations where no other plants could survive.

Exercise 5-D

MUSHROOMS

Kingdom ________________ Phylum/Div. ________________________ Class ______________

Introduction

The mushrooms belong to a group of true fungi that are chiefly saprophytes. They have large conspicuous fruiting bodies above ground and underground hyphae.

Materials Required

1. Mushrooms

Procedure

1. Obtain a mushroom and study its structure.
   A. Note the upper umbrella-like **cap** or **pileus** and **velum.**
   B. On the under surface of the cap observe the **gills.**
   C. Next note the **stalk** or **stipe** which has the root-like **mycelia.**
   D. On the stalk find the **annulus scar.**
   E. Draw and label.

Questions

1. Are edible mushrooms easily identified? Explain.
2. Where are spores formed in mushrooms?
3. Why can mushrooms grow in places other plants cannot?
4. How is the annulus scar formed?
5. Define saprophyte.

UNIT 6

Liverworts, Mosses and Ferns

Exercise 6-A

LIVERWORTS

Kingdom ____________ Phylum/Div. ________________ Class __________

Example—*Marchantia*

Introduction

The Liverworts are somewhat a transition point in the study of plants. This transition is from an aquatic environment to life on land. They live on land, but require water for their reproduction, therefore the transition is not complete.

The Liverworts show considerable advancement over the Fungi and Algae and have multicellular embryos, multicellular sex organs, alternation of generation and live on land.

In this exercise you may use the Liverwort genus *Marchantia.*

Materials Required

1. Preserved specimen (Liverwort)
2. Prepared slides of archegonia, antheridia, and sporangium (Liverwort)
3. Microscope

Procedure

1. First observe the preserved specimens of the Liverwort male and female gametophytes.
   A. The following structures are found on both male and female gametophytes:

      Note the flat body or **thallus**.

      On the lower surface of the thallus find the hair-like structures; these are **rhizoids**.

      On the upper surface of the thallus locate the cuplike structures. They are **gemmae cups** and contain specialized structures which bring about vegetative reproduction.

      On the upper surface of the thallus find the upright **stalks** and observe receptacles at the end.

   B. Determine the sex of your specimen.
      1. The **male receptacle** consists of a **concave disc** with **antheridia inside** (male sex organ).
      2. The **female receptacle** is divided into several **finger-like lobes** with **archegonia underneath**. The archegonia are the female sex organs and are found on the underside of the fingers.

   C. Draw and label each specimen.

2. Obtain a prepared slide of the cross-section of the Liverwort male receptacle showing the antheridia.
   A. Label the **antheridia, the sperm cells,** and the **stalk**.
3. Obtain a prepared slide of the cross-section of the Liverwort female receptacle showing the archegonia.
   A. Observe under the microscope and draw several archegonia.
   B. Label the **jacket cells** which make up the outer parts, the enlarged **base** or **venter**, the **egg** inside the venter and the **neck canal** of an **archegonium**.
4. Obtain a prepared slide of a Liverwort female receptacle bearing sporophytes.
   A. Observe under the microscope and draw.
   B. Label the sporangium **capsule wall** (sheath of cells inside the archegonium) and the clusters or rows of **spores,** the **elaters**, the **seta** and the **foot** (attachment to gametophyte).

Questions

1. How does the Liverwort gametophyte obtain its food?
2. How does the Liverwort sporophyte obtain its food?
3. What is the function of the gemmae cups?
4. Define alternation of generations.
5. Which generation is dominant? Explain why.

Exercise 6-B

MOSS

Kingdom ________________ Phylum/Div. ______________________ Class ______________

Example—*Mnium*

Introduction

The mosses are also transitional plants having alternation of generation, living on land but requiring water for reproduction and bearing multicellular sex organs and multicellular embryos.

Materials Required

1. Preserved specimens of male and female gametophytes.
2. Preserved specimen of female gametophyte bearing a sporophyte.
3. Prepared slides of antheridial and archegonial discs.

Procedure

1. First observe the preserved specimens of the moss male and female gametophytes.
   A. The following structures are found on both male and female plants.
      Rhizoids (rootlike structures at the base of the plants)
      Stemlike axis with small **leaflike appendages** located above the rhizoids.
      Note**These are not true stems and leaves.
   B. The **male disc** is apparent. The **female disc** is hidden by smaller leaflike appendages.
   C. Draw and label.
2. Obtain a preserved specimen of a gametophyte bearing a sporophyte.
   A. Draw and label the structures found in parts A and B above. These make up the gametophyte. The remaining structures make up the sporophyte generation and will not be green in color.
   B. Observe the point of attachment between the gametophyte and the sporophyte. This is the **foot** of the sporophyte.
   C. Connected to the foot will be a long, thin structure, the **stalk** or **seta**.
   D. At the end of the stalk there is an enlarged **sporangium** or **capsule**.
   E. Draw and label.
3. Next obtain a prepared slide of a cross section of an archegonial disc.
   A. Study the slide and locate the following parts of an individual **archegonium**.
      1. It is born on a short **stalk**.
      2. Next is the larger **venter** containing an **egg**.
      3. Extending above the venter is the neck containing the **neck canal**.
   B. Draw and label.

4. Next obtain a prepared slide of a cross section of an antheridial disc.
   A. Study the slide and locate the following structures found on the disc.
      1. The **disc** is attached to the plant by a **stalk**.
      2. The disc contains numerous oval shaped **antheridia** containing **sperm**.
   B. Draw and label.

Questions

1. Define vascular tissue.
2. Do the mosses in this phylum have vascular tissue?
3. Which of the boldfaced structures are diploid? Which are haploid? To which generation does each group belong?
4. What is the function of the sporangium?
5. Define rhizoid.

Exercise 6-C

THE FERN

Kingdom ________________ Phylum/Div. ______________________ Class _____________

Introduction

The fern plant is the first plant to be studied in the laboratory which has true roots, leaves, and stems. These plants have a special vascular system which is used to transport water and dissolved substances from one part of the plant to another. However, these true roots, leaves, and stems are only found in the sporophyte generation.

Materials Required

1. Prepared slides of early and mature prothallus with mature prothallus showing antheridia and archegonia.
2. Frond(fern leaf)
3. Microscope

Procedure

1. First obtain a frond. (Sporophyte leaf)
   A. Note the large flat **blade** and the slender stemlike **petiole**.
   B. Observe the underside of the blade. Along the edge of the blade you will find **sori**, which contain the **sporangia** and produce the spores. They will appear to be light brown in color.
   C. Draw the frond and label.
2. Next carefully remove a single sorus and observe under a microscope.
   A. Carefully break the sorus with a mounting needle and locate a **sporangium**.
   B. Draw and label the **stalk, annulus** (thick-walled cells), and the **spores**.
3. Next obtain a prepared slide of the early prothallus. (Gametophyte)
   A. Draw and label the **prothallus, spore**, and **rhizoids**.
4. Now obtain a prepared slide of the mature prothallus showing the **antheridia** and the **archegonia**.
   A. The antheridia will be found on the underside of the prothallus in the region of the rhizoids.
   B. The archegonia will be found at the opposite end near the notch.
   C. Draw the heart shaped prothallus labeling the **archegonium, antheridium** and the **rhizoids**.

Questions

1. Which generation is dominant in the fern?
2. What is the main difference between rhizoids and rhizomes?
3. Is the gametophyte of the fern monoecious or dioecious?
4. Define or describe a frond.
5. Define or describe a fiddlehead as related to the fern.

UNIT 7

The Pine

Exercise 7-A

THE PINE

Kingdom ________________ Phylum/Div. ______________________ Class ______________

Genus—*Pinus* (Pine)

Introduction

The gymnosperms are seed plants that bear exposed seed as compared to the angiosperms which have flowers that bear their seed enclosed. The gymnosperms are often referred to as "evergreens," that is, plants that do not shed all their leaves at one time. The pine is the example that you will study because it is a typical representative of the gymnosperm plant.

Materials Required

1. Dissecting kit
2. Compound microscope
3. Prepared slide; longitudinal section of male pollen cone of pine
4. Prepared slide; pine pollen (male)
5. Prepared slide; longitudinal section of female ovulate pine cone
6. Prepared slide; pine ovule showing archegonia
7. Pine branches showing first year and second year ovulate cones
8. Pine branches showing mature seed cones
9. Male pine pollen cones
10. First year and second year ovulate pine cones

Procedure

1. Make a drawing of a pollen cone (male) showing the **microsporophyll** arrangement.
2. Observe a microsporophyll from a pollen cone. Locate and label the **microsporangia** or **pollen sacs** on the lower surface.

3. Obtain a prepared slide of pine pollen or prepare a water mount of these by shaking the pollen cone over the glass slide. Identify and label the following:
   A. **Wings**—on either side.
   B. **Prothalli cells**—usually degenerate; found at the curved top.
   C. **Generative cell**—under the prothalli cells.
   D. **Tube cell**—making up most of the pollen grain.
4. Examine a prepared slide of the longitudinal section of a pollen cone (or cut a pollen cone longitudinally) and examine. Identify and label the following parts:
   A. **Microsporophylls**—attached to the axis by a small stalk.
   B. **Microsporangia** or **pollen sacs**—on the lower surface of each microsporophyll.
5. Make a drawing of a first and second year ovulate cone showing the **megasporophyll** arrangement.
6. Observe a prepared slide of a female strobilus (immature ovulate cone). Identify and label the following parts:
   A. **Megasporophyll**—attached to the axis by a small stalk.
   B. **Ovule**—located on top of the megasporophyll.
7. Obtain a prepared slide of a longitudinal section of an ovule. Using low power of the microscope observe the following structures:
   A. **Integument**—outer layer.
   B. **Nucellus**—nutritive portion under the integument.
   C. **Micropyle**—opening at the upper end of integument.
   D. **Megagametophyte**—enlarged central portion of the nucellus.
   E. **Archegonia**—at the micropylar end of megagametophyte.
   F. **Egg**—in each archegonium.

Questions

1. Define angiosperm.
2. Define gymnosperm and coniferous.
3. Explain the difference in pollination and fertilization in the pine?
4. List four commercial uses of the pine.
5. Define annual, biennial, and perennial.

UNIT 8

Roots, Stems, and Leaves

Exercise 8-A

ROOTS

Kingdom ____________ Phylum/Div. ________________ Class __________

Introduction

The prime parts of a plant are the roots, stems, leaves, and flowers. In this exercise we will take up the root system in all of its aspects.

Basically there are two types of root systems, tap root and fibrous or diffused root. The fibrous root system is associated with the monocot plants of which grass is the chief example. The tap root is associated with such plants as legumes.

The tap root system, while not any more elaborate than the diffused root system, has other parts such as a secondary root. Some of the functions of the root system are: absorption, anchorage of the plant, propagation, etc.

Materials Required

1. Specimens of fibrous and tap root
2. Prepared slides of monocot and dicot roots
3. Microscope

Procedure

1. Obtain a specimen of a fibrous and a tap root system.
   A. Observe the **primary** roots on both systems. On the tap root system the single primary root is known as the **tap** root.
   B. Observe and make note of the lateral or **secondary root**.
   C. The terminal end of the root is covered by loosely connected cells making up the **root cap**.
   D. Under the root cap is the **region of meristematic growth**. This is the area of most cell multiplication. (approximately .5 mm)
   E. Moving upward on the root next we have the **region of elongation** or cell specialization. (approximately 1 mm)
   F. The next region of the root is the **region of maturation** where the **root hairs** are located. (approximately 1 mm)
   G. The portion of the root above the maturation region is the **matured region**.
   H. Make drawings of both roots and designate the various areas by labels.

2. This section will deal with the internal structure of the root.
   A. Secure a prepared slide of the monocot plant root. (one seed leaf)
   B. Make a drawing of this specimen and label the following parts: **Xylem**—Usually the largest cells of the vascular system responsible for transportantion of materials up in the plant. **Phloem**—Smaller than the xylem cells, but a part of the vascular bundle. **Pith**—The center area of the root. **Cortex**—Made of parenchyma cells. **Endodermis**—The covering around the vascular bundle. **Epidermis**—The outer covering of the root. **Parenchyma** cells.
   C. Secure a prepared slide of the dicot (2 seed leaf) plant root.
   D. Make a drawing of this specimen and label all the parts which were included in the monocot root with these additions: **Pericyle**—Which is directly below the endodermis and responsible for secondary roots. **Cambium**—Lies between the xylem and phloem cells and gives rise to secondary xylem and phloem. On some of the slides there will be seen secondary roots, if so include these in the drawing.

Questions

1. List four functions of roots.
2. What is the function of the following parts: epidermis, cortex, endodermis, pericycle, phloem, xylem and pith?
3. Define the following: diffuse root system, tap root system, adventitious prop. root.
4. What is the small bulb-like growth on the roots of legume plants which function to store nitrogen?
5. Explain the difference between a rhizoid and a root. A rhizome and a root.

Exercise 8-B

STEMS

Kingdom ________________ Phylum/Div. ______________________ Class ______________

Introduction

The stem of the plant is for upright support and transportation of food materials to the leaves for photosynthesis and the transportation of manufactured food down to the roots where it is stored.

Stems vary in many shapes, sizes, and description; ranging from herbaceous to woody in structure.

Materials Required

1. Specimen of woody and herbaceous stems
2. Prepared slides of herbaceous stems (monocot and dicot)
3. Prepared slide of a woody stem

Procedure

1. Make a drawing of a herbaceous stem (grass).
   - A. Note from what specific area the leaves arise.
   - B. Note if more than one leaf arises from the same general area.
2. Make a drawing of a woody stem.
   - A. The following parts should be located by use of the textbook and chart. **Lenticel, terminal bud, terminal bud scar, node, internode, bud** (if present).
   - B. Observe where a leaf has been removed and locate the **bundle scar**, and **pith**.
3. Obtain a prepared slide of a woody stem. (Dicot—2 seed leaf)
   - A. The inner most part is the **pith**.
   - B. The **annual rings** are dead xylem cells. Only the last ring of xylem cells, for the most part, are functional.
   - C. Note the **cambium** and **phloem cells**. It is well worth mentioning that the phloem cells (only under rare instances will this not be true) are closer to the outside than the **xylem cells**.
   - D. The epidermis is composed of **cork cells** which are usually quite thick in comparison to the other cells.
   - E. Prepare a drawing of the woody stem (again refer to text for guide) cross section labeling the following parts: **pith, xylem cells, phloem cells, vascular rays, cambium,** and **epidermis**.
4. Obtain a prepared slide of a monocot. (1 seed leaf)
   - A. Using the monocot slide, which is often called the monkey face vascular bundle, find and be able to identify the following parts: **vascular bundle, xylem, phloem, bundle sheath, epidermis, parenchyma cells,** and **pith**.
   - B. Prepare a drawing of the monocot cross section and label.

5. Obtain a prepared slide of a dicot stem. (Herbaceous)
   A. Note the vascular bundle of the dicot stem forms a ring near the outer edge of the stem itself. This should aid you in recognizing the dicot stem and not confusing it with the monocot stem.
   B. Using the prepared slide of the dicot stem find and be able to identify the following parts: **vascular bundle, xylem, phloem, bundle sheath, epidermis, parenchyma cells, cambium,** and **pith**. Prepare a drawing, labeling the parts.
   C. It is well worth noting here there are many companion cells in the vascular bundle of a stem. However, it will be unlikely you will see these, but possible.

Questions

1. Give the function of the following: vascular bundle, xylem, phloem, cambium, epidermis, parenchyma cells and sclerenchyma cells.
2. How is an annual ring formed in a tree?
3. Define or give the function of the following: node, internode, lenticel, bud, terminal bud and apical meristem.
4. What does the distance between the terminal bud and last year's terminal bud scar represent?
5. Define or describe heartwood and sapwood.

Exercise 8-C

LEAVES

Kingdom ________________ Phylum/Div. ______________________ Class ____________

Introduction

Each part of the plant functions in its own right. Carrying on the bulk of the photosynthetic process, the leaf produces glucose which is stored as starch. Any green area of the plant is capable of photosynthesis, but due to the usual concentration of chlorophyll in the leaves, this makes them the greater producer.

Photosynthesis is the second greatest chemical change or reaction on earth, with evaporation and condensation being the greatest.

One of the easiest ways of determining whether a plant is a monocot or dicot is by the venation of the leaf. If the plant has parallel venation it is a monocot. If the plant has netted venation, then it is a dicot.

Materials Required

1. Leaves to examine and determine the margin types (optional)
2. Prepared slide of leaf epidermis
3. Prepared slides of the leaf cross section
4. Microscope

Procedure

1. We begin this exercise with an examination of the internal structure of the leaf.
   A. Obtain a prepared slide and locate the following parts working from the upper surface of the leaf down: **cuticle, upper epidermis, palisade cells, mesophyll cells, xylem** and **phloem, stomata, guard cells,** and **lower epidermis**.
   B. By use of the prepared slide draw and label.
   C. Obtain a slide of the lower epidermis of a leaf. Label **stomata** and **guard cells**.

Questions

1. What is the prime function of the leaf?
2. What is the function of the following: stomata, palisade cells, mesophyll cells, cuticle and guard cells?
3. Explain the guard cells function.
4. Define deciduous.
5. Define cotyledon.

UNIT 9

The Flower

Exercise 9-A

FLOWER

Kingdom ________________ Phylum/Div. ______________________ Class ______________

Introduction

The flower, like the leaves, stem, and root, is found on both monocot and dicot plants.

All organisms, both animal and plant, must reproduce their own species if the individual is to survive. The flower of a plant serves this function. Some flowers have only the male reproductive parts with the female reproductive parts on another flower. Many flowers have both male and female parts. As a general rule of thumb there is very little self fertilization.

This exercise is designed to acquaint the student with the parts of the flower and the function of the various parts.

Materials Required

1. Fresh flower
2. Prepared slides of pollen grains
3. Prepared slides of developed ovary
4. Dissecting kit
5. Bi-nocular microscope

Procedure

1. The first section will deal with the fresh flower.
   A. Follow along with your instructor and dissect the flower.
   B. The following will be the step by step process of tearing the flower apart in order to determine where the various parts are to be found. (Label on plate 4)
      1. **Receptacle**—Expanded part of the stem.
      2. **Calyx**—All **sepals** (Sepals—leaflike parts)
      3. **Corolla**—All **petals** (Petal—delicately textured and often brightly colored)
      4. **Stamen**—Male (Includes **anther** and **filament**)
      5. **Pistil**—Female (Includes **stigma, style** and **ovary**)

2. After the flower has been completely dissected and all of the parts have been identified make a cross section cut in the ovary and observe the locules (small pocket-like objects in the ovary).
3. Using a prepared slide of a pollen grain observe a **pollen grain** using high power. Label it and its **nuclei** on plate 4.
4. Use a prepared slide of an ovary, locate and draw the following cells in their proper place: **egg, synergid, antipodals, polar nuclei**. Refer to the diagram for their location (Plate 4).

Questions

1. List all the male parts of a flower.
2. List all the female parts of the flower.
3. List all the accessory parts of a flower.
4. Explain the terms perfect and complete as they relate to a flower.
5. Flower color results from what two pigment groups?
6. What is self fertilization?
7. What is cross fertilization?
8. Why is self fertilization harmful?
9. What is the function of the anther?
10. What is the receptacle?

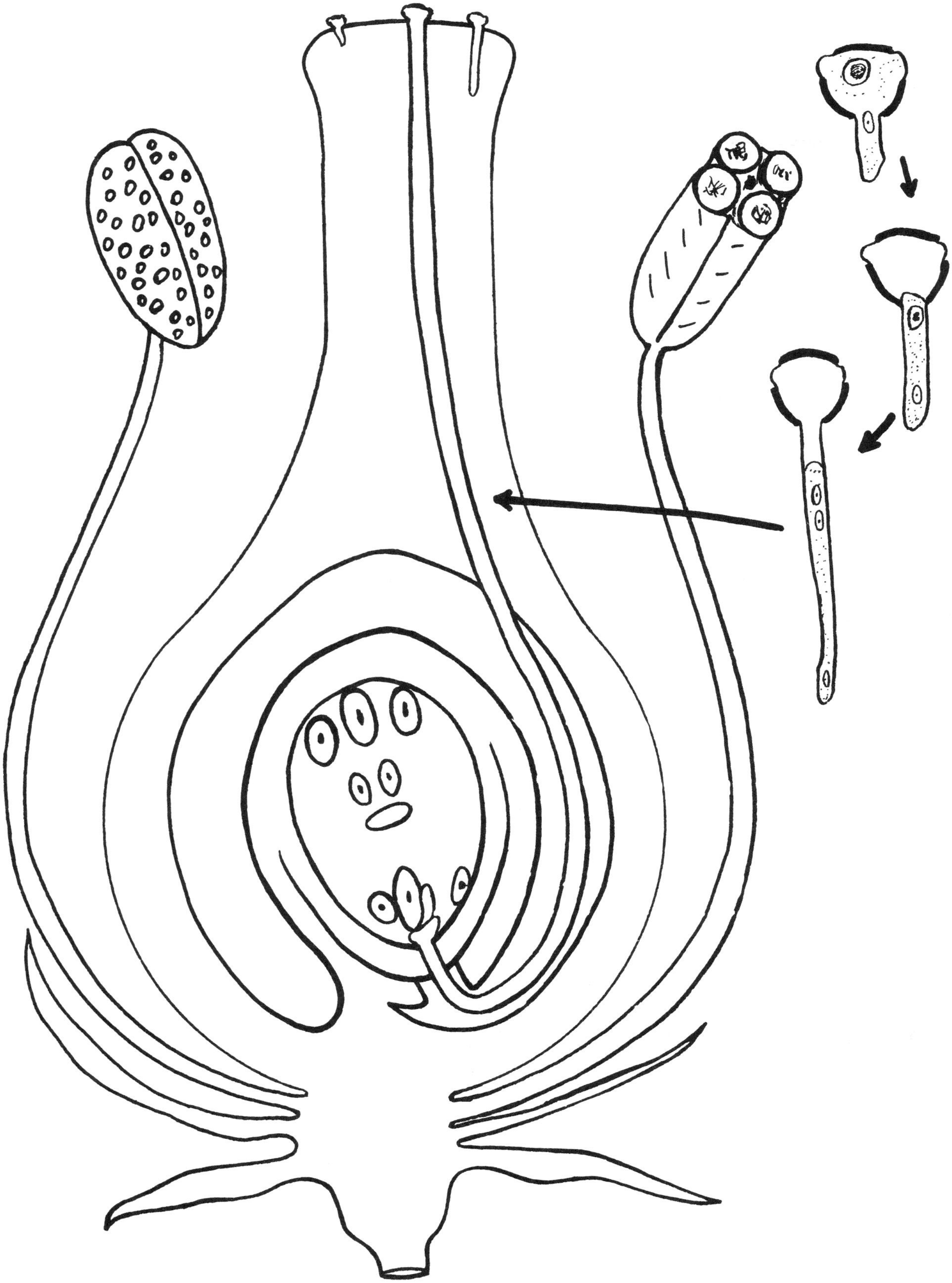

Plate 4. The Flower

Exercise 11-B

PARAMECIUM BEHAVIOR

Introduction

The PARAMECIUM is easily cultivated by placing dead grass, or hay, in a container and covering with water. After this mixture has been allowed to incubate for a period of seven (7) to ten (10) days large numbers of PARAMECIA will have been cultivated.

Materials Required

1. Clean glass slides and cover glasses
2. Paramecium culture
3. Congo red dye (a pH indicator)
4. Yeast culture (yeast cells that have been boiled and stained with congo red dye)
5. Acetic acid
6. Blue ink
7. Methyl cellulose (thickening agent)

Procedure

1. PARAMECIUM reaction to acid.
   A. Place a drop of PARAMECIUM culture on a clean glass slide and cover with a cover glass. Place the slide on your microscope and observe movement of the organism under both low and high powers.
      1. Describe the degree of flexibility and the motion of the body as it travels through the liquid.
      2. How many contractile vacuoles are there in the PARAMECIUM?
      3. Where are the contractile vacuoles located?
      4. Do the contractile vacuoles move through the body as food vacuoles do, or do they stay in one place?
      5. How many contractions are there of a contractile vacuole in a period of one (1) minute?

   B. Focus your microscope on 10X and place a small drop of acetic acid at the edge of the cover glass. Observe the direction of movement and note any changes in patterns of movement as the acid diffuses through the water.
      1. What is the initial reaction of the organisms to the acetic acid as it diffuses through the liquid?
      2. Is there any further change in their reaction? Describe.
2. Demonstration of the PARAMECIUM feeding response.
   A. Place a drop of PARAMECIUM culture on a clean glass slide. Add a drop of the congo red yeast culture and cover with a cover glass.

   B. Isolate a single PARAMECIUM under high power.
      1. Describe the movement of cells near the surface of the PARAMECIUM.
      2. Describe the movement of the cells near the oral groove.

3. Where do the food vacuoles form?
4. Describe the movement and color change of the cells as the food vacuole leaves its site of formation.

3. PARAMECIUM exploding its trichocysts.
   A. Place a drop of PARAMECIUM culture on a clean glass slide and cover with a cover glass. Focus under low power on your microscope.
   B. Place a drop of ordinary blue ink at the edge of the cover glass. Observe the reactions of the organisms as the ink diffuses through the water.
      1. Make a drawing of a PARAMECIUM and label the **trichocysts** as they are seen around the organism.

Sponge, Hydra, and Jellyfish

Exercise 12-A

SPONGES

Kingdom ________________ Phylum/Div. ______________________ Class ______________

Example: *Scypha*

Introduction

Sponges represent the most primitive multicellular animals. Sponges are sedentary and exhibit no localized movements, but they are far more active than they appear. Some sponges may draw in and expel up to 50 gallons of water in a 24-hour period. The specimen you will study is *Scypha,* a marine form that lives in clusters in shallow water, where it is usually attached to rocks.

Materials Required

1. Preserved *Scypha* sponges
2. Prepared slide sponge, longitudinal section
3. Hand lenses or dissecting microscope

Procedure

1. External Structure: Locate the following structures on a preserved specimen.
   A. **Osculum**—opening at end of the sponge.
   B. **Spicules**—stiff, rodlike structures all over the surface of the sponge.
   C. **Base**—the closed end of the sponge.
   D. **Dermal ostia**—tiny pores on surface which open into central cavity.
   E. **Buds**—small projections near base of the animal (your specimen may not have these).
2. Make a longitudinal cut through the midline of the body from the osculum to base with a sharp razor blade or scalpel. Place the two halves in a container with water and examine the **central cavity** or **spongocoel** with hand lens or dissecting microscope. Find the small internal ostia or **apopyles** which open from the **radial canals** into the central cavity. Try to find the tiny canals in the walls of the sponge. Note the **spicules** in the walls. Draw and label.
3. Internal Structure: Obtain a prepared slide of a longitudinal section of sponge. Locate the following structures and label on plate 5.
   A. **Central cavity** or **Spongocoel**—in the middle of the section.
   B. **Radial canals** or **Excurrent canal**—which open into the central cavity or spongocoel.
   C. **Incurrent canals**—canal from the outside.

D. **Prosopyle**—opening from external surface or incurrent canal.
E. **Apopyle**—exit of the excurrent canal.
F. **Choanocytes**—also called collar cells, line the inner surface of the excurrent canals.
G. **Epithelium**—lining of external surface, incurrent canals, and lining of the spongocoel.

Questions

1. Define marine.
2. Are all sponges marine? Explain your answer.
3. List 4 uses of sponges.
4. Explain how sponges are harvested.
5. Explain why sponges are referred to as the filters of the sea.

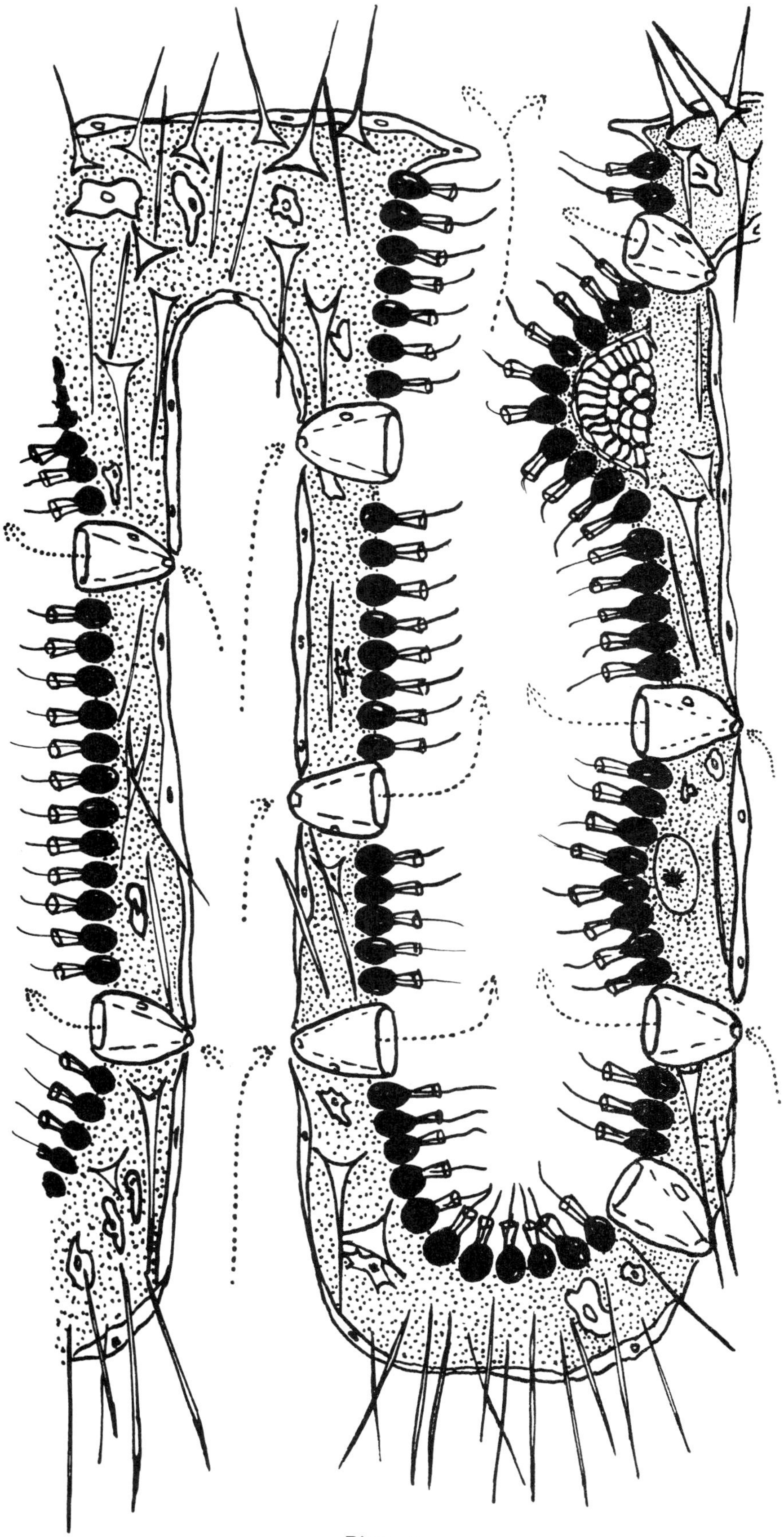

Plate 5.

Exercise 12-C

GONIONEMUS
(medusa or jellyfish form)

Kingdom _______________ Phylum/Div. ______________________ Class ____________

Introduction

Gonionemus is frequently studied as a type of jellyfish, since the medusa form is much larger than most other coelenterates. It is commonly found in shallow marine water in certain regions of the eastern coast of the United States.

Materials Required

1. Preserved *Gonionemus*
2. Watch glasses or other shallow dishes such as petri dishes.
3. Hand lenses or dissecting microscopes.

Procedure

1. Obtain a preserved specimen and place in a shallow dish filled with water.
2. The jellyfish is very fragile, so use care in handling it.
3. Examine it with hand lens or under dissecting microscope.
4. Draw a side view and a bottom view.
5. Label the following structures on the view indicated (S = side, B = Bottom, Both = Both drawings)
   - A. Outside surface of the bell—**exumbrella.**(S)
   - B. Inside or **oral surface** of the bell which is concave—**subumbrella** (hangs outside the ventral surface). (B)
   - C. **Tentacles**—fringe-like projections from the margin of exumbrella. (Both)
   - D. **Stinging cells**—clusters on tiny rings of swelling area on tentacles. (Both)
   - E. **Adhesive pad**—sharp bend near distal end of tentacles. (Both)
   - F. **Tentacular bulbs** or **ocelli**—pigmented area at base of tentacles. (Both)
   - G. **Statocysts**—tiny structures located between the bases of the tentacles. (B)
   - H.. **Velum**—circular shelf-like membrane; forms the outer margin of the subumbrella cavity. (B)
   - I. **Manubrium**—structure hanging down from inside of bell; much as the handle of an ordinary umbrella. (Both)
   - J. **Mouth**—at distant end of manubrium with four lip-like oral lobes. (Both)
   - K. **Gullet**—leads from mouth to the stomach. (Both)
   - L. **Stomach**—which is at base of the manubrium. (Both)
   - M. **Radial canals**—four canals extending from the stomach region to the margin of the medusa. (Both)
   - N. **Circular or Ring canal**—connects with the radial canals; at the margin of the medusa. (Both)

O. **Gonads** —irregular, convoluted structures; suspended under each of the radial canals. (B)
P. **Epidermis**—exumbrella surface, subumbrella surface and appendages. (S)
Q. **Gastrodermis**—lines the gastrovascular cavity and the radial and circular canals. (S)
R. **Mesoglea**—middle jelly-like layer and makes up a greater percentage of the organism. (S)

Questions

1. If you should come in contact with a jellyfish while swimming this could be a painful experience. Explain why.
2. Where would you have to be swimming to come into contact with a jellyfish?
3. How does the jellyfish move?
4. Why is the jellyfish always tilted in the direction of movement?
5. Is jet propulsion the only method of movement in jellyfish? Explain your answer.
6. What is the function of the statocysts in jellyfish movement?
7. List 2 characteristics of jellyfish that are also found in the hydra.
8. Define gonad.
9. Since jellyfish can swim, why do you see so many washed up on beaches?
10. Explain alternation of generation in jellyfish. Which stage or generation did you study in lab?

Exercise 14-B

TRICHINELLA

Kingdom ________________ Phylum/Div. ____________________ Class ____________

Genus—*Trichinella*

Species—*Spiralis*

Introduction

Trichinella spiralis is a round worm and is responsible for the serious disease trichinosis. The adults live in the small intestines where the female burrows into the intestinal wall and produces larvae (small worm), which burrow into veins or lymph vessels and are carried to the skeleton muscles. In the muscles they coil and are surrounded with a limy cyst. In the encysted stage they may live for years.

Materials Required

1. Prepared slide of *Trichinella spiralis.*
2. Microscope.

Procedure

1. Obtain a prepared slide and focus under 10× objective.
2. Find small larvae coiled up in the muscle fibers of the meat.
3. The larvae are found in small pockets known as cysts.
4. Locate a cyst in which a complete coil is visible. Draw the **muscle fibers, larva** and the **cyst.**

Questions

1. List 4 symptoms of trichinosis.
2. Explain how you become infected with trichinosis.
3. How do you prevent trichinosis infestation?
4. List the 2 most common types of meat that may be infested with trichinella.
5. It has been estimated that approximately 15% of the population of the United States has been infested with trichinella at some time or other in their life. Explain the reason for this.

The Crayfish and Grasshopper

Exercise 16-A

THE CRAYFISH

Kingdom ________________ Phylum/Div. ______________________ Class ____________

Introduction

The phylum *Arthropoda* is the largest group of animals on earth. More than three-fourths of all kinds of animals on earth belong to this group; which indicates that this group is the most successful on earth. As man's chief competitor the arthropods are of great economic importance. The arthropods are joint-footed animals (Greek: arthon—joint+pous—footed).

Materials Required

1. Preserved crayfish
2. Dissecting pan
3. Dissecting kit

Procedure

Part I. External Anatomy

A. Obtain a preserved specimen of the crayfish and a dissecting pan and study the external features as follows: Label the following structures on plate 10.
    1. Note the numerous appendages, i.e., pinchers (chelae), legs, "feelers" attached to the body. The body is divided into two main body regions, (a) the **cephalothorax,** the rigid anterior half of the body covered by a hard shell, the **carapace,** and the (b) **abdomen,** the flexible segmented posterior half of the body. Observe the **cervical groove,** a groove in the carapace. The **head** is anterior to the groove on the cephalothorax and the **thorax** is just behind the groove.
    2. Examine the **tail** and its five parts. The central part at the tip is the **telson.** Located on the ventral surface of the telson is the **anus**.
    3. Identify the **eyes,** located on stalks. These stalks can be drawn into the head.
    4. Located on the anterior part of the head are several "feelers." The two long "feelers" are **antennae** and the smaller are **antennules.**
    5. Below the base of each antennae are found the **excretory pores.**
    6. Label the above structures on plate 10.

Part II. Appendages

A. Appendages of the thorax and abdomen. (Label the following on plate 10.)

1. Locate the **chelae** (pinchers) and the four pair of **walking legs** attached to the thorax.
2. The first five pair of appendages on the abdomen are called the **swimmeretts.**
3. The last segment on the abdomen is the **telson.** This segment has no appendages. On the next to last segment are the uropods. The fan-shaped **uropods** are flat plates.

B. The remaining appendages that we will study are thoracic and head appendages. Follow directions carefully in removal of the appendages to insure proper identification.

1. Before removal, observe a demonstration of the properly removed appendages.
2. Tear out plate 10 showing the appendages.
3. As you remove the appendages from the left side of the specimen by clipping through the point of attachment; place the antenna, antennule, chela, and four walking legs on the plate in proper order.
4. The remaining appendages are to be removed from the "mouth" region. The following appendages occur from the outermost third maxilliped to the innermost mandible:
   a. **3rd maxilliped**—Touch, taste, handle food.
   b. **2nd maxilliped**—Touch, taste, handle food.
   c. **1st maxilliped**—Touch, taste, handle food.
   d. **2nd maxilla**—Draws water into the gills.
   e. **1st maxilla**—Handles food.
   f. **Mandible**—Crushes food.
5. To remove the appendages, find the base of the left third maxilliped; grasp firmly at the base and pull loose. You will observe that the appendages are similar to the pages of a book in relation to the position as they lie on top of one another. With great care, remove the remaining appendages in the same manner as you did the last, placing the appendages on the paper. There should be thirteen appendages. Ask your lab assistant to check your removal before you proceed with the remaining exercise.

Part III. Determine the sex of your crayfish as follows: In the **female** look for a large, triangular cavity between the bases of the third and fourth pairs of walking legs. This structure is the **seminal receptacle.** Located at the base of the **second pair of walking legs** are the two **oviduct openings.** (Look for a circular opening, covered by a thin membrane, at the base of each appendage.) The **male** is recognized by having the **first two pairs of abdominal appendages** pressed together and extending between the base of walking legs. These appendages are modified to form the **copulatory organ** which transfers sperm to the seminal receptacle of the female. At the base of the **fourth pair of walking legs** in the male, find the **sperm duct openings.** Draw a ventral view of each sex.

Part IV. Internal Anatomy (Label all the boldface structures on plate 11.)

A. Observe a demonstration of the proper method of dissection and then proceed.

1. With your scissors, cut, beginning on the carapace-abdomen joint, anteriorly to a point just behind the eyes. Then, make a cut behind each eye. Remove the halves of the carapace carefully to expose the internal structures.
2. To remove the shell from the dorsal side of the abdomen, insert the scissors under the first segment about ¼ inch to the side of the midline and cut backwards to the last segment. Do the same on the other side to remove the shell. The **extensor muscle of the abdomen** will probably be also removed. The renaining large mass of muscle is the **flexor muscle of the abdomen.**

B. Identify the following structures:
   1. **Gills**—Numerous feathery structures on the side of the animal in the thorax region.
   2. **Heart**—A rectangular structure in the midline of the dorsal portion of the thorax just anterior to the abdomen.
   3. **Ostia**—Small openings in the heart.
   4. Remove the gills from the left side of the crayfish, carefully removing the thin transparent sheet which covers the internal organs.
   5. **Digestive gland**—A large, cream colored lobe, consisting of two lobes, one on each side of the animal.
   6. **Ovaries or testes**—(Depending on sex) They lie directly under the heart, between the digestive glands. The **testes** are two whitish lobes extending to the abdomen. Extending from the testes are coiled tubes, the **sperm ducts,** which carry sperm to the outside. The **ovaries** are in the same position, but the oviducts are lacking.
   7. **Intestine**—Located in the depression of the flexor muscle. It extends to the stomach.

C. Remove the digestive gland on the left side to expose further structures.
   1. **Gastric mill—(Stomach)** A large structure in the front part of the thorax. You will note two rows of internal bars used in grinding of food.
   2. **Esophagus**—A short thick tube connecting the mouth to the stomach.
   3. **Green glands**—Two structures on either side of the esophagus and connected to the excretory pores at the base of the antennae.

D. Nervous System
   1. By removal of the digestive gland, heart, gills, intestine and stomach the nervous system will be exposed. Identify these parts:
      a. **Circumesophageal connectives**—Two nerves, extending anteriorly around the esophagus and connected to the brain.
      b. **Brain**—A concentration of nerve tissue in the head region. Note the branches.
   2. Expose the ventral nerve cord by removing the large flexor muscle from the abdomen.
      a. Identify the **ventral nerve cord** and observe the swellings found on it; these are the **ganglia.**
      b. **Subesophageal ganglion**—Posterior to the esophagus. It gives rise to the **circumesophageal connectives** which pass around the gullet to the brain.

Questions

1. How are the gills and circulatory systems related in reference to oxygen-carbon dioxide exchange?
2. Does the crayfish have a closed or open circulatory system?
3. When does growth occur in the crayfish? Explain.
4. Define: chitin, hemocoel, ganglion.
5. Define: autotomy, regeneration, cannibalism.

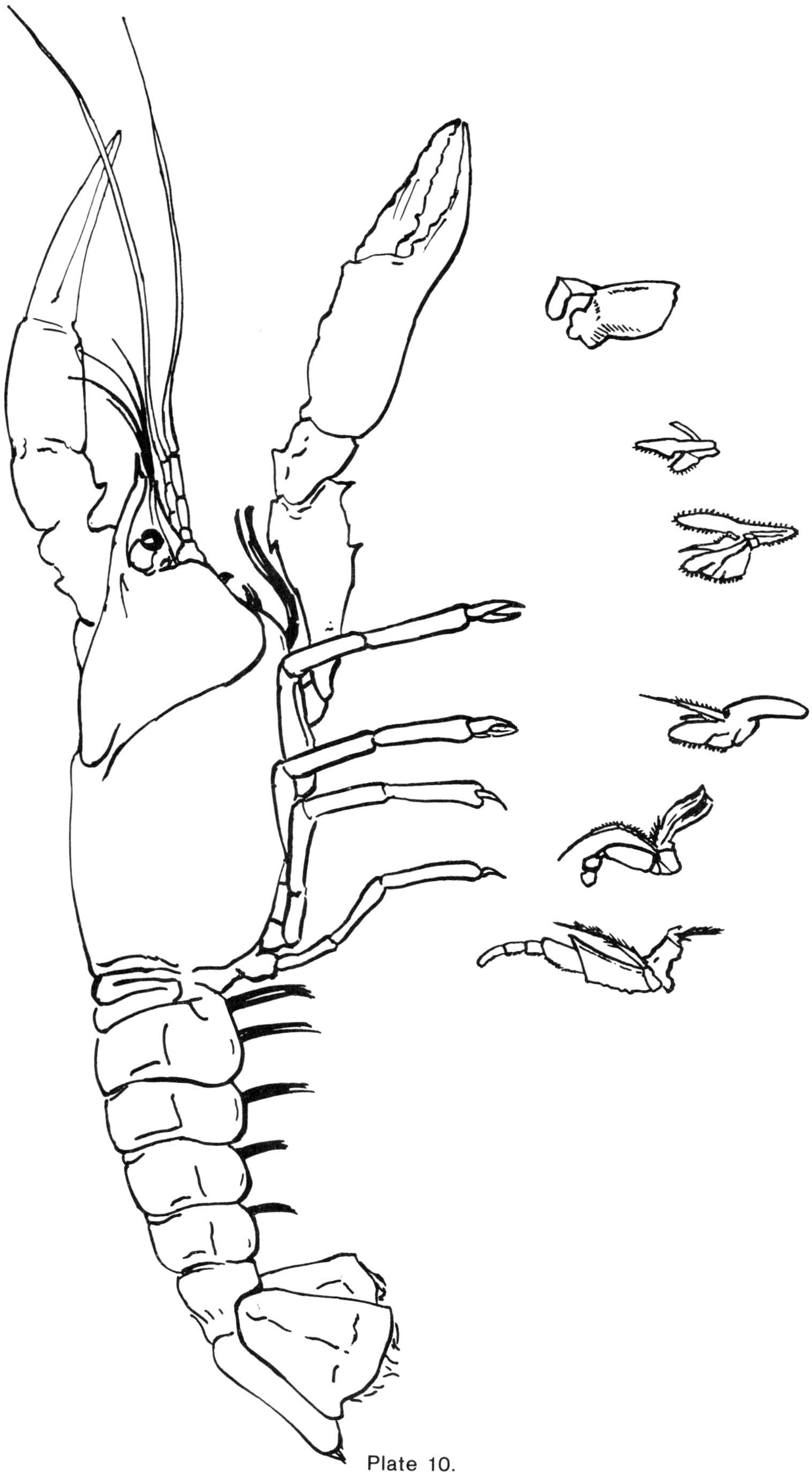

Plate 10.

16-5

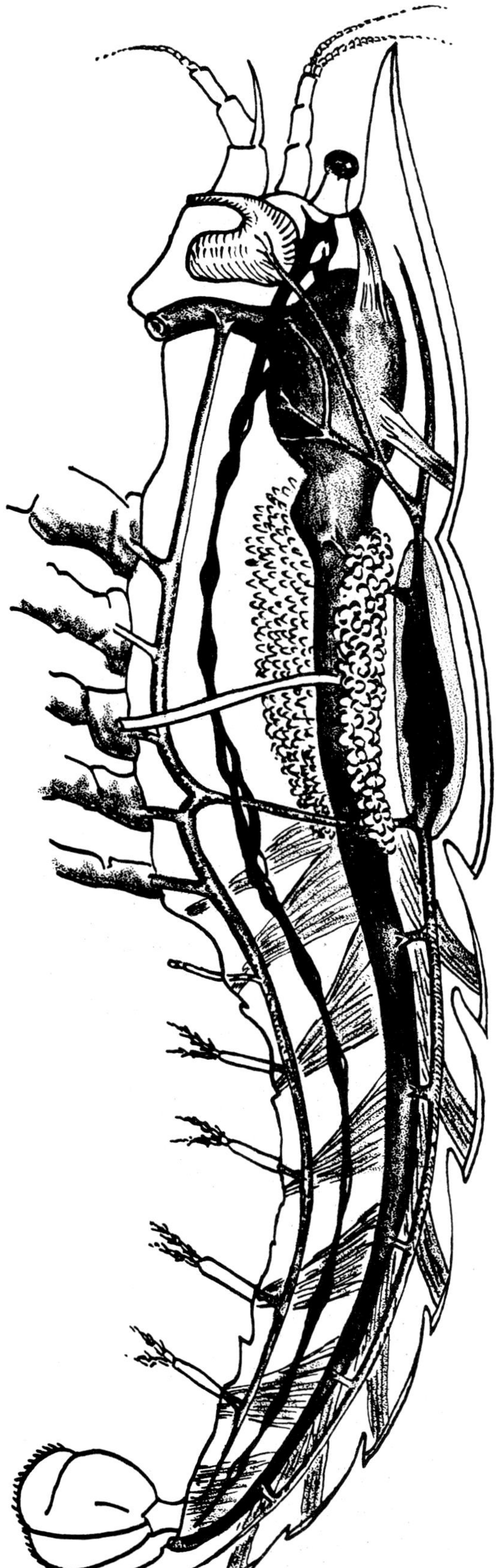

Plate 11.

UNIT 17

The Starfish

Exercise 17-A

STARFISH

Kingdom ________________ Phylum/Div. ______________________ Class ______________

Introduction

These are the "spinny skinned" animals of the oceans. They are abundant in the ocean primarily because they seem to have very few natural enemies. Most members of this phylum, with a few exceptions, rely on a water vascular system for locomotion. The species you will study is the starfish.

Materials Required

1. Dissecting pan
2. Dissecting kit
3. Preserved starfish
4. Slide: cross section of starfish ray (optional)

Procedure

External structure:

1. Secure a preserved specimen and identify the following structures.
   A. **Central disc**—central portion of animal.
   B. **Rays** (arms)—that project from the central disc.
   C. **Aboral surface**—surface opposite the mouth.
   D. **Oral surface**—mouth side
   E. **Mouth**—opening in center of oral side.
   F. **Madreporite**—button like structure between the junction of two rays.
   G. **Bivium**—the two rays nearest the madreporite.
   H. **Trivium**—the other rays extending from the disc.
2. Study the following surface structures.
   A. **Spines**—distributed over the surface of the animal.
   B. **Pedicellariae**—around the base of the each spine raised ring of skin bearing small pincer like structures. (Use hand lens or dissecting microscope to see these.)
   C. **Skin gills**—between spines; soft, pale finger like projections.

3. On the oral surface find the following structures.
   A. **Mouth**—surrounded by 5 pair of oral spines.
   B. **Ambulacral grooves**
   C. **Tube feet**—finger like projections.
   D. **Suction cups**—little indentations at end of each tube foot.
4. Make two drawings: One showing the oral surface; the other showing other structures on surface of starfish: Label.
5. Internal structure (dissection)

   Place your specimen under water in the dissecting pan with the oral surface downward. Using scissors, cut off about ½″ of the terminal end of a trivium. Using extreme care do not damage underlying tissues and organs, cut through the body wall along the lateral margin of the ray. Lift the terminal end of the ray and carefully free any membranes clinging to the body wall by using the dull side of the scalpel. Identify and label the following structures.

   A. Digestive system:
      1. **Mouth**—center of oral surface.
      2. **Stomach**—above mouth.
      3. **Intestine**—short, from stomach to anus.
      4. **Anus**—on aboral surface.
      5. **Hepatic caeca**—one pair in each ray; greenish, long, glandular, and lobed.
      6. **Rectal caeca**—two, at base of intestine; short lobes.
   B. Reproductive system:
      1. **Gonads**—two lobed in rays below the hepatic caeca.
      2. **Reproductive duct**—small tube leading from gonad to the genital pore.
      3. **Genital pore**—conspicuous opening on aboral surface.
   C. Water vascular system: parts of this system are difficult to see.
      1. **Madreporite**—external opening on aboral surface.
      2. **Stone canal**—curved tube from madreporite to ring canal.
      3. **Ring canal**—circular tube around mouth region.
      4. **Radial canal**—one in each ray, extending from the ring canal.
      5. **Transverse canal**—from radial canal to ampullae; short.
      6. **Ampullae**—in floor of coelom; spherical, connect through pore to the tube feet.

Questions

1. What are pedicellariae and what is their function?
2. Describe respiration, excretion, and circulation in the starfish.
3. What is meant by autotomy?
4. Why is it a bad policy to chop up starfish and throw them back in the water if you are an oyster fisherman?
5. What function do the tube feet of a starfish serve?
6. Describe the feeding habits of the starfish.

The Clam

Exercise 18-A

THE CLAM

Kingdom ________________ Phylum/Div. ______________________ Class ______________

Introduction

The *Mollusks* have a soft body generally with an anterior head and a ventral foot. The body is surrounded by a thin mantle and is commonly enclosed in an external covering or shell.

In this exercise we will study the clam or mussel.

Materials Required

1. Preserved clam
2. Dissecting kit
3. Dissecting pan

Procedure

1. Obtain a preserved specimen and study the external features.
   A. The oval shell is an exoskeleton and is composed of two parts, the **right** and **left valves**.
      1. The valves are connected by a **hinge** on the dorsal region. The dorsal region is the thicker martins of the valves.
      2. The oldest portion of the shell is located on the anterior and may appear to be a hump on each valve. This is known as the **umbo**.
      3. Note the **growth lines** around the umbo.
      4. The wide bands on the valves show about one year's growth.
      5. Draw and label the external features.
2. Internal anatomy of the clam:
   A. Remove the left valve and study the inside surface.
      1. Find the scars on the valve left by the muscles.
      2. On the dorsal anterior portion of the valve note the large scar left by the **anterior adductor muscle**. At the dorsal posterior edge of this scar find the scar left by the **anterior retractor muscle**.
      3. On the dorsal posterior portion of the valve find the large scar from the **posterior adductor muscle**. Dorsal to this scar and continuous with it find the scar from the **posterior retractor muscle**.

4. Draw the inside of this valve and label the scars.

3. Next study the body of the clam as it lies on the right valve with the ventral anterior surface toward you.
   A. Note that the body is enclosed in a soft membrane. This is the mantle, and conforms to the valves. These two sides are joined to form a cavity known as the mantle cavity.
      1. Now study the posterior end of the mantle and note the two openings. The larger posterior opening is the **incurrent siphon** and it is surrounded by finger like **palps**. The smaller ventral opening is the **excurrent siphon**.
      2. Note the attachment of the mantle to the valve.
      3. Carefully remove the left portion of the mantle to expose the **mantle cavity**.
      4. Locate the **large muscular foot**. It is found in the anterior ventral portion of the body.
      5. Next locate the **gills** posterior and slightly dorsal to the foot.
      6. Between the foot and the anterior adductor muscle locate the slit that forms the **mouth** which is surrounded by two small **labial palps**.
      7. Underneath the gills you will see the **visceral hump** which contains the internal organs.
      8. The **anus** is located at the posterior end of the visceral hump near the dorsal surface.
      9. Draw and label the exposed body of the clam.

Questions

1. What is the function of the adductor muscle?
2. What is the function of the retractor muscle?
3. What is the function of the incurrent siphon?
4. What is the function of the excurrent siphon?
5. What are two functions of the exoskeleton?

The Amphioxus and Perch

Exercise 19-A

AMPHIOXUS

Kingdom ____________ Phylum/Div. ________________ Class __________

Introduction

The phylum Chordata is the best known phylum because it includes the largest and most conspicuous animals in the animal kingdom. All of the Chordate members are alike since each one has at some stage of its life a rod-like notochord for support, a hollow, dorsal nerve cord, and gill slits.

Branchiostoma lanceolatus, commonly known as amphioxus, lives in waters of tropical and temperate regions. The amphioxus is important since it exhibits the characteristics of the chordates in a simple condition.

Amphioxus is about two inches long with a semitransparent body, pointed at both ends and laterally compressed. It is found near the shore, burrowed into the sand with only the feeding or anterior end protruding.

Materials Required

1. Prepared slide of amphioxus
2. Microscope

External Anatomy

1. The amphioxus is shaped like a fish but has no lateral fins and no distinct head.
2. Along the middorsal line is a low **dorsal fin** which extends the entire length of the body and widens at the posterior end into a **caudal fin**.
3. The caudal fin extends forward on the ventral surface to form the short ventral fin.
4. The dorsal and ventral fins are supported by rods call **fin rays**.
5. In front of the ventral fin, the ventral portion of the body is flattened with two expansions called the **metapleural fold**.
6. The **body wall** is divided into V-shaped muscle segments, the **myotomes**. They produce lateral movement of the body when swimming.
7. The **mouth** is an opening in a membrane posterior to the **oral hood**.
8. The **anus** is situated on the left side of the body near the base of the anal fin.
9. The **atriopore**, through which water passes to the outside in respiration is located anterior to the ventral fin.

Internal Anatomy

1. Located near the dorsal surface at the body and nearly extending the body length find the rod-like **notochord**.
2. Part of the skeletal structures support the **oral tentacles**, projections of the oral hood.
3. The mouth is an opening in a membrane, the **velum**. Projecting posterior from the velum, locate the velar tentacles. The tentacles act as a strainer for food materials.
4. Posterior to the mouth is the **pharynx** with slit-like openings called **gill slits**. Separating the openings are **gill bars**.
5. Water with food enters the mouth, pharynx and passes through the gill slits into the atrium, a cavity and finally to the outside through the atriopore.
6. Extending backward from the pharynx is the intestine and a "pocket," the **hepatic diverticulum**, extending anteriorly from the intestine.
7. Lying above the notochord, find the **central nerve cord**.
8. Find the **gonads**, which appear to extend below the hepatic diverticulum. There are two sexes and the sperm and eggs are discharged through the atriopore. Fertilization then takes place in the water.
9. Label on plate 14.

Plate 14.

Exercise 19-B

THE PERCH

Kingdom ____________ Phylum/Div. ________________ Class ____________

Introduction

The bony fish represents the largest group of vertebrates both in number and species and in number of individual. They are found in both fresh and salt water.

Materials Required

1. Preserved perch
2. Stained slides of fish blood (if available)

Procedure

1. External structure—Obtain specimen and identify the following.
   A. **Head**—extending to posterior end of operculum.
   B. **Trunk**—extending to anus.
   C. **Tail**—to the end of the caudal fin.
   D. **Operculum**—covering over gills.
   E. Identify the fins:
      1. **Pectoral fin**—posterior and slightly dorsal to operculum.
      2. **Pelvic fin**—posterior and ventral to operculum.
      3. **Anal fin**—behind anal opening on ventral surface.
      4. **Dorsal fin**—on dorsal surface.
      5. **Caudal fin**—tail fin.
   F. **External nares** (nostrils)—Put a needle into a nare to see if this connects to mouth cavity.
   G. **Eyes**—Examine them to see if they have lids.
   H. **Nictitating membrane**—thin membrane covering eyes.
   I. **Gills**—lift up operculum to examine these.
   J. **Lateral line**—row of small pores in the scales running from the center of the base of the caudal fin to the operculum.
   K. **Scales**—Remove one of the scales of the lateral line and examine under a microscope: note the small pore.
   L. **Anus**—on ventral surface anterior to anal fin.
   M. **Urogenital papilla** (opening)—slitlike, small depression posterior to anus.
   N. **Mouth**—at anterior portion of head.
   O. Draw and label external structures on drawing paper.

2. Internal anatomy: Identify the following structures and label on plate 15.

   Obtain a specimen of the perch. Place it in dissecting pan and cover with water. Using scissors make a cut from the anus to the head in the mid-ventral line being careful not to damage underlying organs. Make a horizontal cut on one side from the mid-ventral incision to the back bone at the level of the origin of the anal fin. Make a similar cut at the base of the pectoral fin. Remove the flap of the body wall carefully exposing the internal organs.

   Identify the following structures:

   A. **Liver**—anterior end on ventral surface; somewhat lobed.
   B. **Stomach**—elongated.
   C. **Intestine**—which is usually surrounded by yellow fat.
   D. **Gonads**—in the posterior part of the cavity.
   E. **Air bladder**—lies dorsal to these organs and is thin walled.
   F. Digestive system:
      1. Examine **mouth** and **pharynx**.
      2. Run probe into the opening of the **esophagus** at the posterior end of the pharynx.
      3. Look under the **liver** and trace the esophagus to the **cardiac** portion of the stomach (anterior, larger portion).
      4. The **pyloric** portion extends ventrally and connects to the small intestine.
      5. **Pyloric caeca**—left side of cardiac pouch; three of these.
      6. **Intestine**—extends to the exterior opening, the anus.
      7. **Spleen**—grayish; between first loop of intestine and the blind end of the stomach.
      8. **Gall bladder**—under the right lobe of the liver.
      9. **Common bile duct**—from the gall bladder to the intestine at the base of the pyloric caeca.
   G. Reproductive system:
      1. Each sex is separate and is difficult to distinguish externally.
      2. **Ovary** (if female)—saclike organ located above the intestine and continues posteriorly toward the urogenital pore as a tube, **ovarian duct**.
      3. If you have a male perch, the **testes** can be seen as paired structures just above the coils of the intestine.
      4. **Vas deferens**—extend from the posterior end of each testis to the **urogenital pore**.
   H. Respiratory and circulatory system:
      1. Open the mouth of the perch and make a horizontal cut from the angle of the jaw through the operculum and gills into the esophagus.
      2. **Teeth**—on upper and lower jaw.
      3. **Tongue**—on ventral surface of the mouth.
      4. **Oral flaps**—thin flaps of tissue behind the teeth on both jaws.
      5. **Gill arches**—bars which support the other breathing structures.
      6. Cut away one gill arch, place in water, and examine with hand lens or dissecting microscope. Identify the following: **Gill filaments**—double row on outer side of the arch. **Gill rakers**—inner surface of each arch to strain out food organisms.
      7. **Heart**—located in front of the liver and underneath the back of the pharynx.

I. Muscular system:

1. Remove the skin from the right side of the body (or part of the side that is not cut away) and expose the muscles underneath.

   A. **Myotomes**—zigzag bands of segmental muscles.

   B. **Horizontal septum**—separates the myotomes: This is commonly known as connective tissue.

Questions

1. What is the function of the air bladder?
2. What is the function of the lateral line?
3. What forms the lateral line?
4. Compare the skeletal system of the perch and shark.
5. What is the function of the nictitating membrane?
6. What organ manufactures bile?
7. What organ stores and secretes bile?
8. What is the function of bile?
9. What is the function of the gills? Explain your answer.
10. Which fin functions in propulsion: and what fins function in guidance?

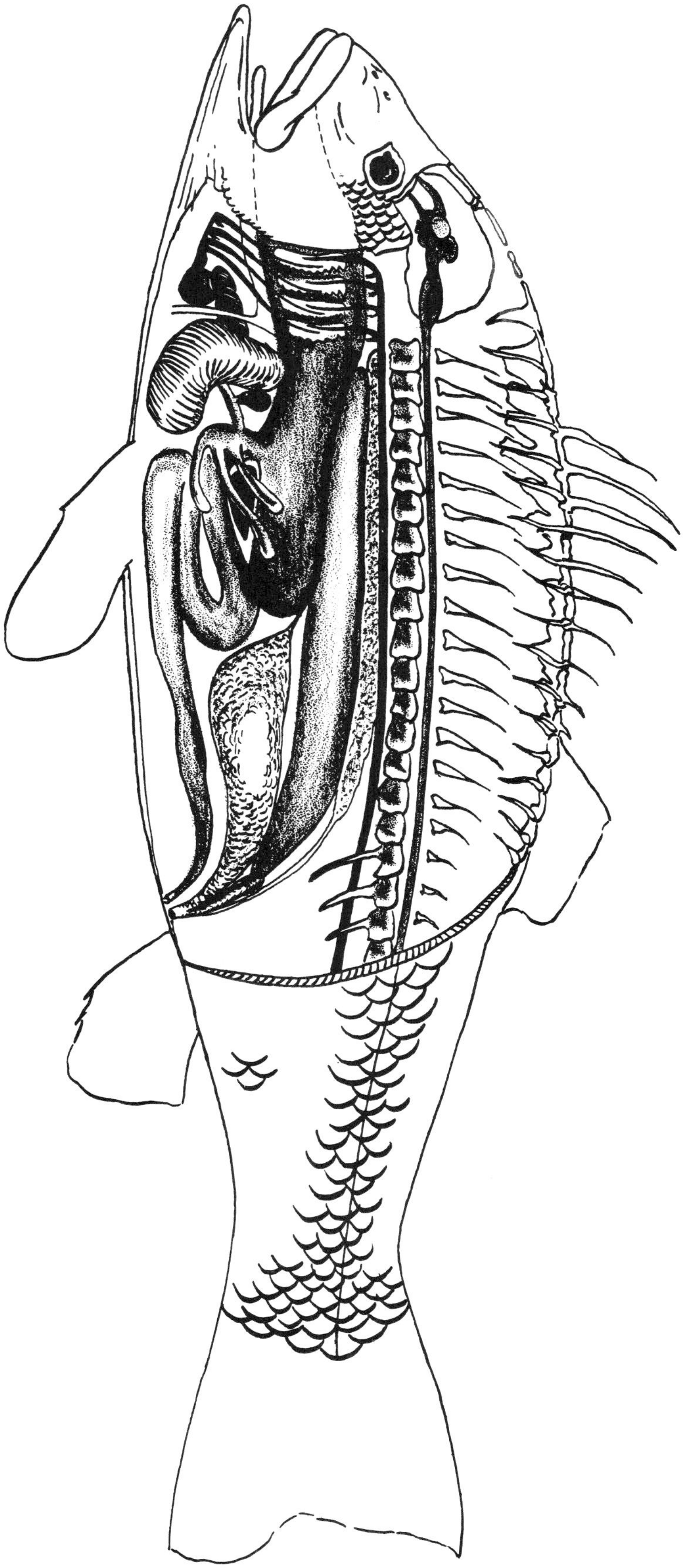

Plate 15.

UNIT 20

The Frog

Exercise 20-A

THE FROG

Kingdom ________________ Phylum/Div. ______________________ Class ____________

Introduction

The frog is somewhat a transition point in the study of animals. In the early stages of development (tadpole stage) the frog lives in an aquatic environment due to the fact that he must receive oxygen by use of gills. Developing into an adult, the frog develops lungs and is able to extract some oxygen from the human's normal environment. The frog still maintains dependence upon the aquatic environment by returning frequently.

In this exercise you may use one or both of the following species of frogs: *Rana pipiens*—the leopard or grass frog, *Rana catesbeiana*—the common bullfrog.

Although this animal makes the transition from water to land he does not lose his aquatic characteristics and spends much time in the water for lubrication of the external body parts, as well as receiving some oxygen through the skin from the water.

In this exercise you will become familiar with both the internal and external parts of the frog's body. It is of the utmost importance you learn this lesson well as its application will be made to the human body.

Materials Required

1. A living specimen of the frog.
2. Dissecting kit.
3. Dissecting pan.
4. Textbook.
5. Mounted charts.

Procedure

(Living specimen) (for the most part the following will also apply to a preserved specimen)

1. Observe the external anatomy of the frog. Find and identify the following parts: **mouth, nostrils (external nares), eyes, eyelids, tympanic membrane, head, trunk, forelegs, hindlegs (jumping legs), toes, anus, dorsal surface, ventral surface, right lateral, left lateral, anterior** and **posterior ends** and **skin.**

A. Draw the external anatomy of the frog and label.

B. Note the thickness of the skin. It is through this covering many of the respiratory gases are exchanged.

2. Dissection—Spread the frog out in a dissection pan and pin the body to the wax bottom with the ventral side up. Lift the loose skin of the frog and make the incision from the posterior to the anterior end of your specimen. Make lateral cuts in front of the jumping legs and behind the fore legs. Pull the skin back and pin. Note: Hold the skin up as you make your cuts being careful not to cut into any muscle.

   Expose the body cavity by cutting through the ventral body wall (carefully along one side of the midline). Cut through the bones of the **shoulder girdle,** make lateral cuts and pin this skin back. In cutting through the body wall you have cut the **somatic peritoneum,** a thin membrane that forms the outer lining of the **coelom.** The body organs lie in the coelom.

   Next locate and observe the organs within the coelom using a blunt instrument. Only on rare occasions will the scissors be used once inside the body cavity. Note the **heart** with the major vessels leading both away from and to the heart. Note the **lungs** (two sac like structures) located slightly anterior and ventral to the heart. Look on both sides and posterior to the heart and observe the **liver** (the largest organ in the body). Lying between the right and middle lobes of the liver is the greenish **gall bladder** which stores bile. Under the left lobe of the liver you will find the **stomach**. A short **esophagus** connects the mouth and the stomach. The **duodenum** leads from the stomach through the **pyloric sphincter** valve into the **small intestine** (a small coiled tube) which enlarges to become the **large intestine** which terminates into the **rectum** which voids outside through the **cloaca** and **anal opening.** The **spleen** is a deep red, spherical organ embedded in the mesentery supporting the small intestine. The **pancreas** is somewhat pink, lobed and lies near the liver. The **urinary bladder** is ventral to the large intestine and is nearly transparent. LABEL THE ABOVE PARTS ON PLATE 16.

3. Locate the reproductive organs of the MALE frog. Note the yellow bands on the **kidneys.** These are the **adrenal glands.** To locate the **testes** look just above the kidneys on both sides of the **spinal column.** They will appear as two small, whitish, egg shaped structures. Trace the **duct** which leads from the testis to the cloaca, which serves as a passage for both waste material and sperm. LABEL THE ABOVE PARTS ON PLATE 17 of your lab manual.

4. Locate the reproductive organs of the FEMALE frog. Since you will only dissect one frog, it will be necessary for you to swap your frog for one of the opposite sex. Note the **ovaries** which may be heavy with **eggs**, (black spotted masses lying on both sides of the body cavity). The ovaries are connected to a long greatly curved, **oviduct** which runs almost the entire length of the body cavity. Note the **ostium** (opening) of the oviduct at the anterior end of the body cavity. Trace the path of the oviduct which empties into the **cloaca.** LABEL THE ABOVE PARTS ON PLATE 18 of your lab manual.

5. Dissect your frog exposing the following bones: **nasal, front parietal, carpals, radioulna, humerus, thoracic vertebra, pubis, ischium, tarsus, phalanges, metatarsals, tibiofibula, femur, ileum, urostyle, sacral vertebra, lumbar vertebra, scapula, clavicle, metacarpals, phalanges** on fore-legs, and **maxilla.** LABEL THE PARTS ON PLATE 19 of your lab manual.

6. From your preserved specimen dissect out the **brain** and be able to identify the following parts: **olfactory lobe, cerebrum, optic lobe, medulla oblongata, cerebellum,** and **spinal cord.**
   Draw a picture of the brain and its parts. Use the parts above as your labels.

Questions

1. Explain why the frog is classified as an amphibian.
2. Compare the frog's heart with the heart of man.
3. Why is the frog said to be cold blooded?
4. List 5 functions of the frog's liver.
5. Explain how to tell a male from a female without dissecting the specimen. List at least two (2) ways.

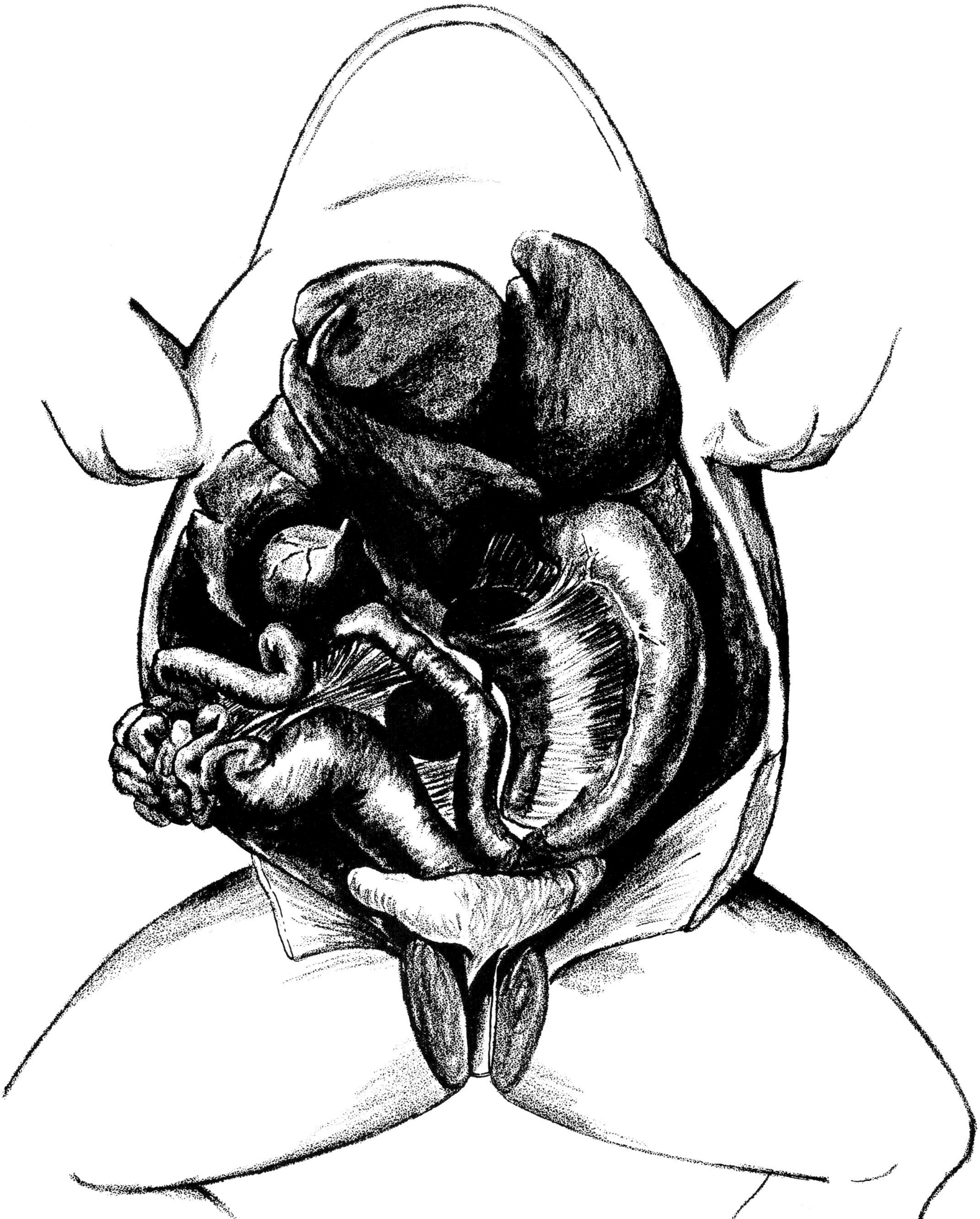

Plate 16.

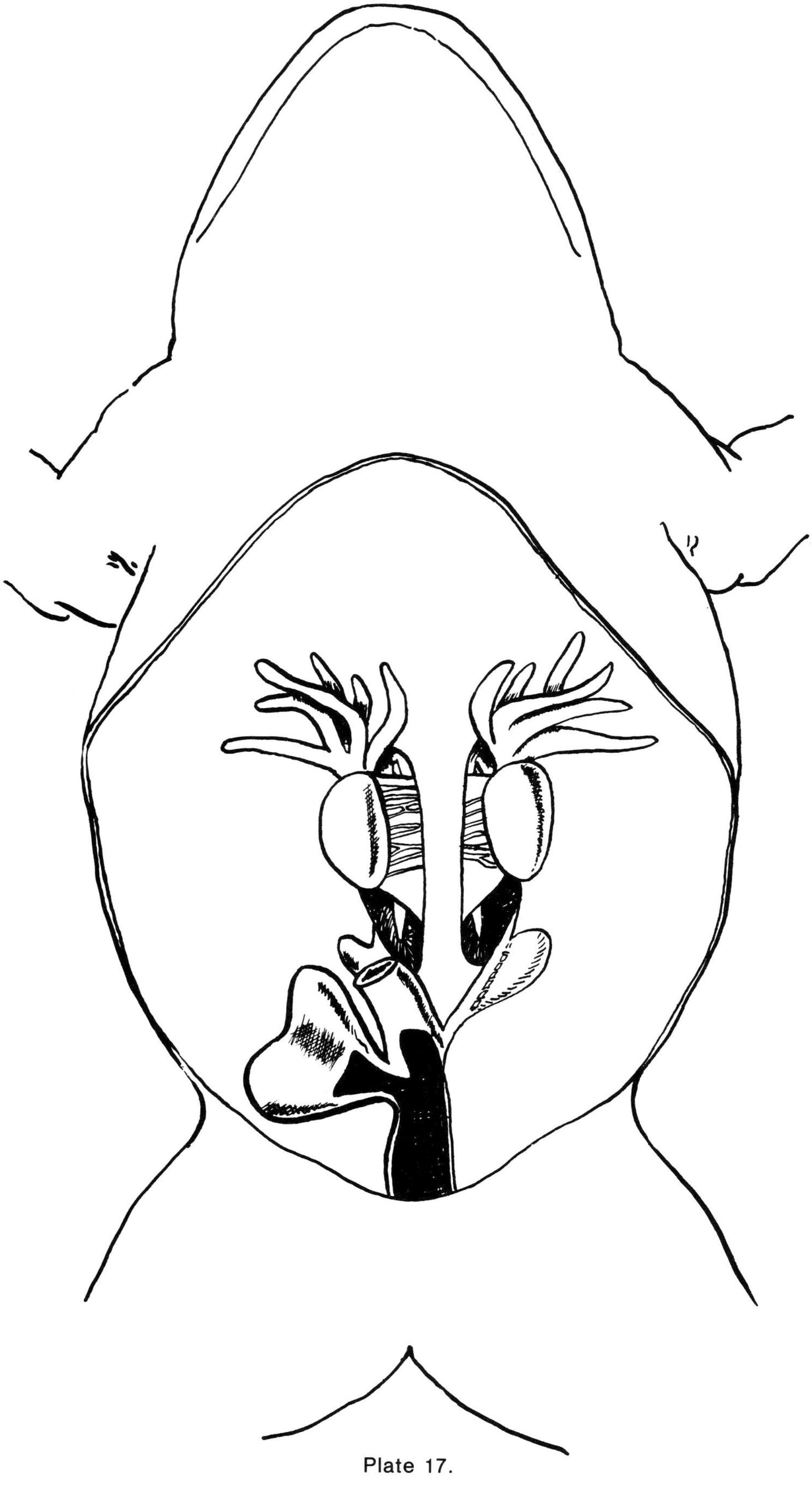

Plate 17.

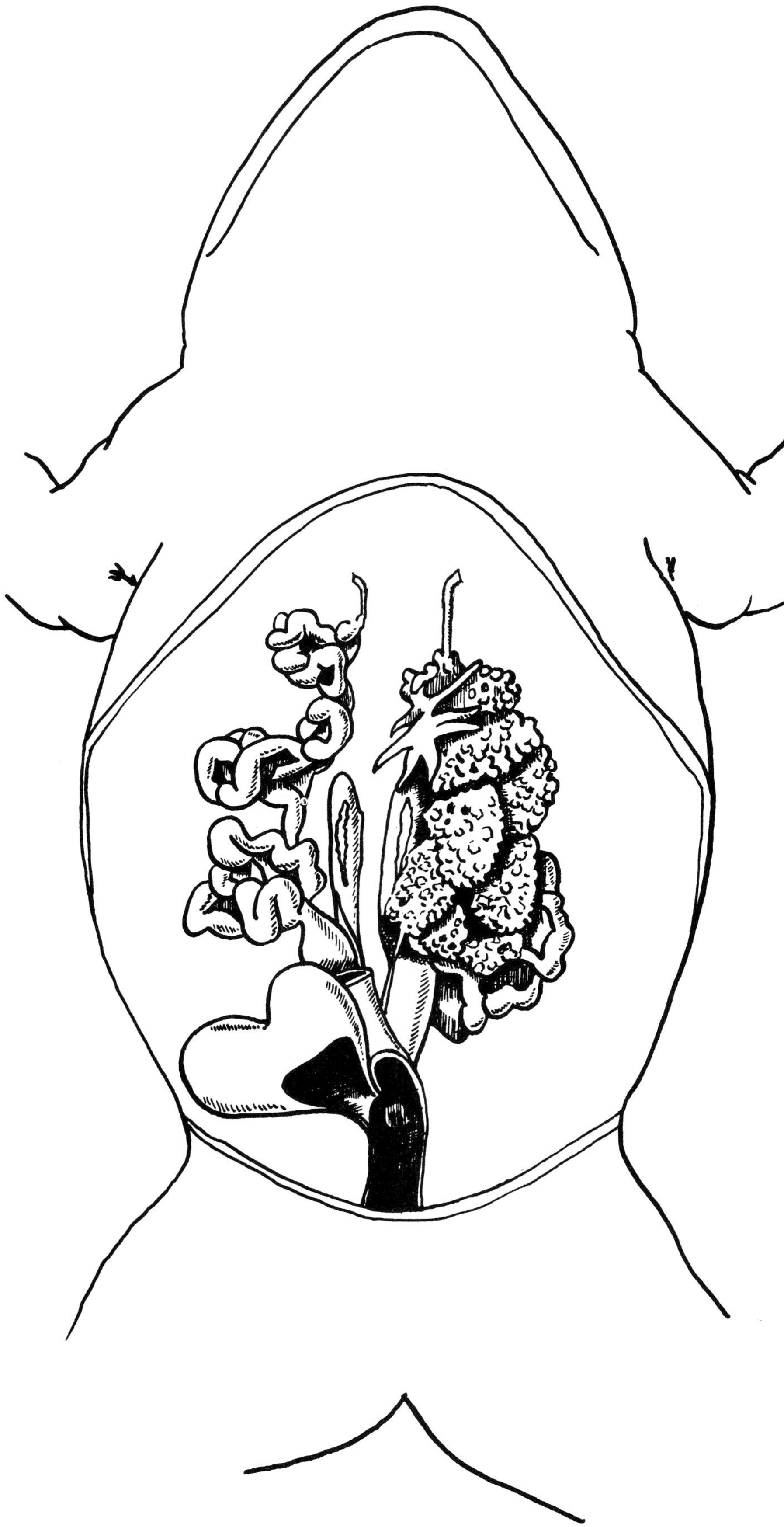

Plate 18.

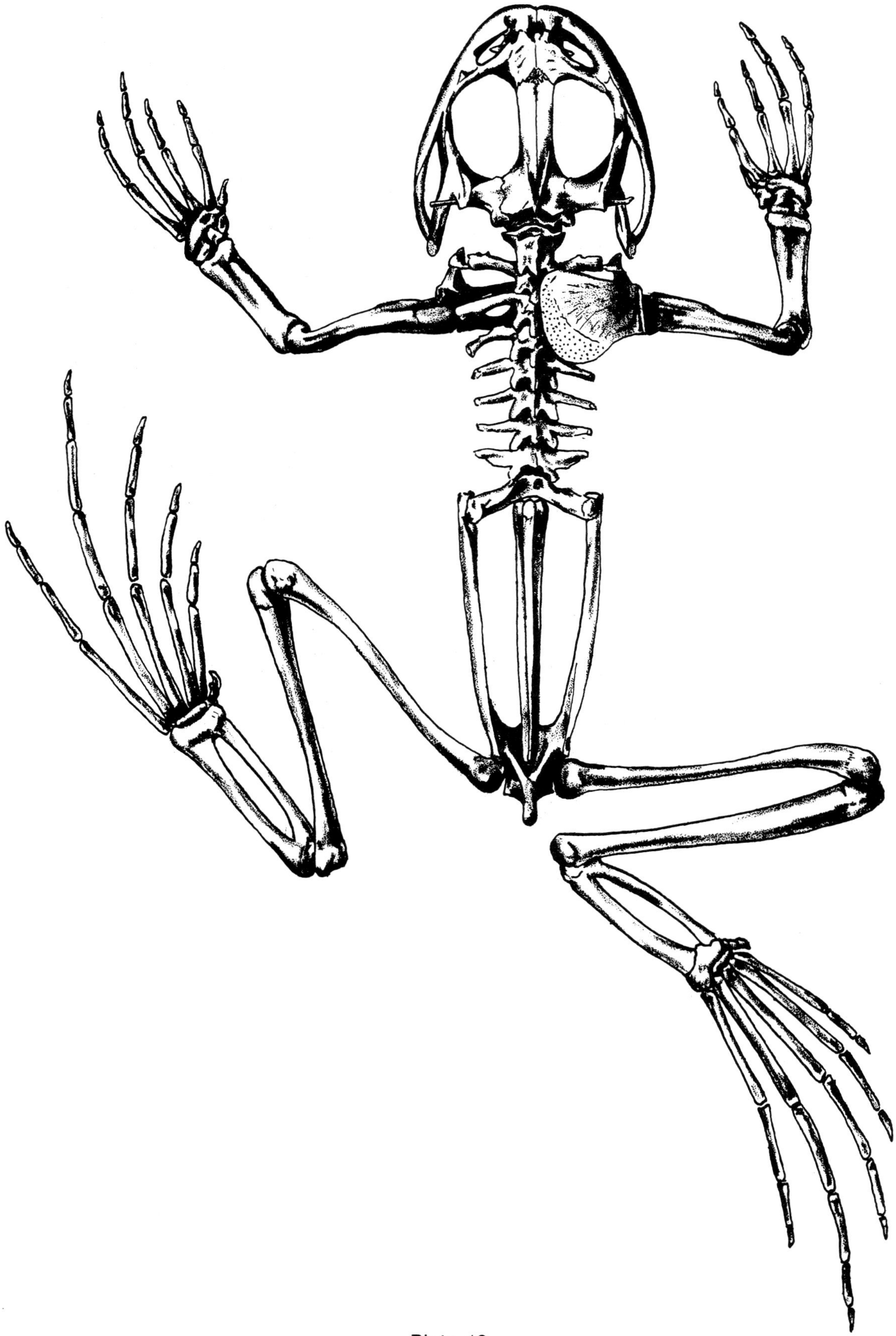

Plate 19.

UNIT 21

The Rat and Pig

Exercise 21-A

RAT

Kingdom ________________ Phylum/Div. ______________________ Class ______________

Introduction

In the study of zoology it is necessary that at least one mammal be used to partially complete the coverage of the animal kingdom. In this exercise we will use the rat as the representative animal of the mammals.

The rat, though a lower form of animal, has many characteristics associated with the human. Although a rat is a rat basicly, the reproductive cycle is the same as in man.

Materials Required

1. Preserved or anesthetized rat
2. Dissecting kit
3. Dissecting pan

Procedure

1. If your lab is to use a live animal it is necessary that it first be anesthetized. This is accomplished by placing a pad of ether or chloroform soaked cotton in a large mouth jar. The rat is then placed in the container with the anesthetic and allowed to remain until unconscious. If ether is used it will probably be necessary to keep a saturated pad lying on the nostrils in order to keep the animal asleep. Of course, if using a preserved specimen, this will not be necessary.
2. Place the specimen in the dissecting pan with the ventral side up. By use of a sharpened scalpel make an incision into the body cavity just immediately anterior to a line between the hind legs. By using the scissors complete the incision anteriorly to a line just posterior to the forelegs.
3. After this is accomplished make lateral cuts and pin the skin to the dissecting pan. Again by use of the scissors cut the rib cage at the cartilage union of the sternum and ribs proper, cutting well up into the esophagus area. This should now have exposed the internal organs of the body.

   By use of text, chart (rat and human), preserved specimen dissected accomplish the following. If your specimen is a living specimen make careful note of the action of the heart and the flow of blood into the aorta. Locate and be able to identify the following portions of the internal structure: **esophagus, trachea, aorta, heart, lungs, liver, stomach, pancreas, spleen, kidney, large intestine, mesenter, small intestine, uterus, rectum, vagina,** and **bladder.** (Refer to plate 20). Fully label. Next turn the rat with the dorsal side up in order to allow removal of the skull and provide an entrance to the brain.

Make a lateral incision posterior to the skull, then make two anteriorly directed cuts and "scalp" the organism. By use of heavy surgeon scissors and scalpel make an incision in the skull completely around the top. Continue this work very delicately until the brain is exposed.

Make an incision at the junction of the spinal cord and the brain and cut posteriorly so the entire spinal column is exposed. By further dissection remove the brain and spinal cord in tact if possible. Make notation of the oblongata, cerebrum and other parts of the nervous system.

Questions

1. List several characteristics of a mammal.
2. Trace the reproductive cycle, by words, beginning with the sperm and egg development until the zygote is formed.
3. List at least 10 characteristics which the rat has that is related to the human.
4. List the parts of the internal structures to be labeled and give the function of each part.

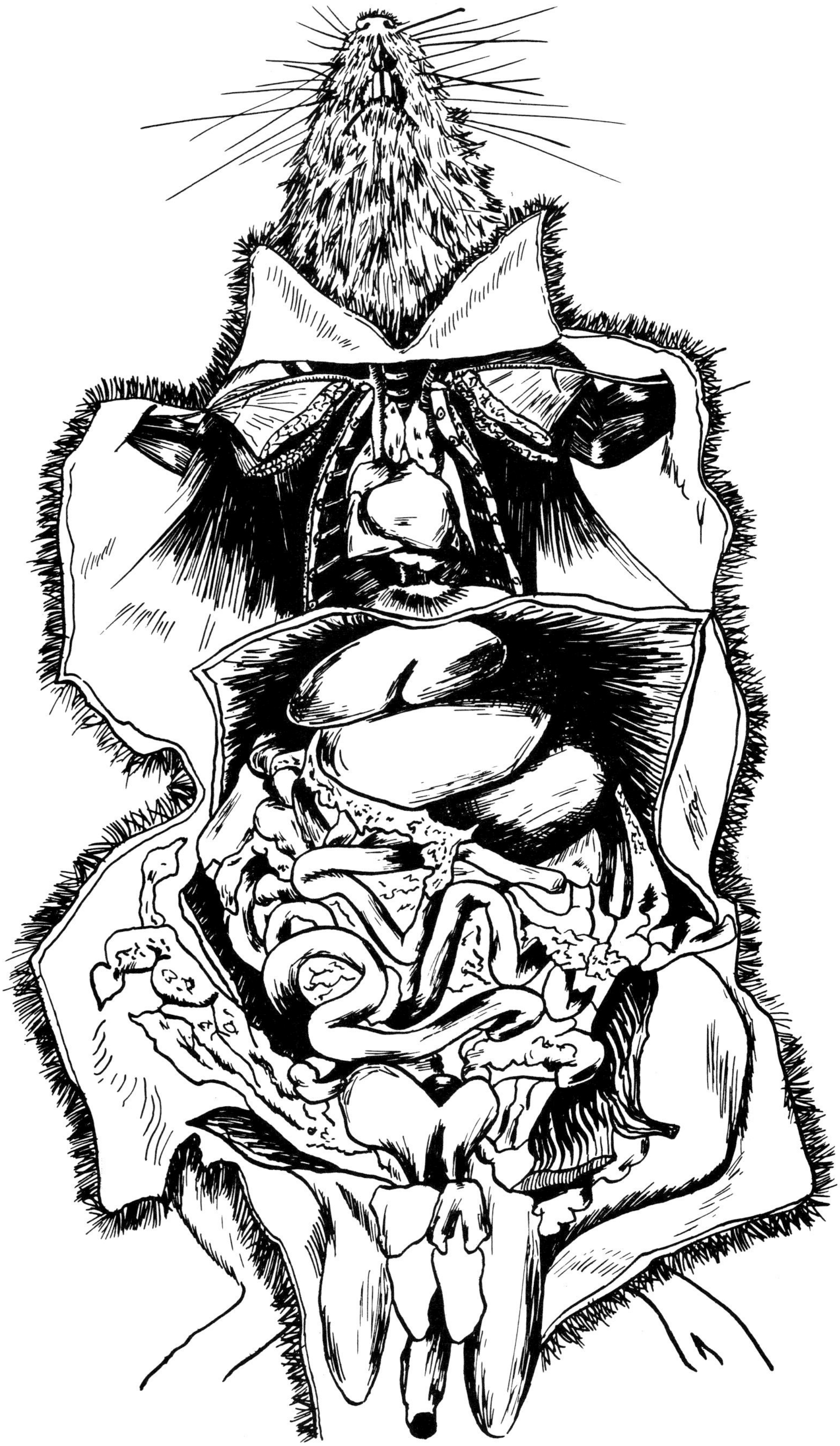

Plate 20.

Exercise 21-B

FETAL PIG

Kingdom ____________________ Phylum/Div. ____________________________ Class ________________

Introduction

Pigs closely resemble humans, anatomically and physiologically. Pigs, like ourselves are omnivores, so their digestive tract are similar to ours. As a result, in this lab you will learn about animal structure and function by examination of the fetal pig.

Materials Required

1. Preserved fetal pig
2. Dissecting kit
3. String
4. Dissecting pan

Procedure

1. (EXTERNAL ANATOMY)–At the sink remove the fetal pig from the storage container or plastic bag and rinse it off under the tap. Do not be alarmed at the appearance of your specimen. Remember that mammals, including humans, are born dripping wet since they spend the fetal period in a watery environment. Notice the body of a mammal is divided into a number of regions. The head is separated from the trunk by a moveable neck. The trunk is generally divided into the anterior thorax and posterior abdomen and pelvis. Most mammals have a tail.
   A. Examine the head and locate the eyes, the external ears, or pinnae (singular: pinna), the nostrils and the mouth.
   B. Notice the paired legs and their various joints.
   C. Compare the pig forelimb with your own arm, and try to figure out the corresponding parts.
   D. Do the same for the hindlimb.
   E. Pick up the pig and manipulate the joints. Do they move the same way yours move?
   F. Notice the toes. How many are there on each foot? ____________________________
   G. How many toes does the pig actually walk on? ____________________________
   The pig walks on the tip of their toenails, a method of locomotion known as ''unguligrade.'' How does this differ from the human method of walking? ____________________________
   ____
   H. Observe the **umbilical cord** by which the fetal pig was attached to the placenta during development within the uterus.

I. Is your pig male or female? _________ Raise the tail and observe the **anus,** the posterior end of the digestive tract. If your specimen is a female, there will be a second opening just ventral to the anus, the urogenital opening. This is the exit for both the reproductive and excretory system. Protruding from the urogenital opening in the female is a small, fleshy urogenital papilla. If your specimen is male, there is no second opening near the anus. Instead, locate the small **urogenital** opening just posterior to the umbilical cord. The duct leading to it runs forward from between the legs in a long muscular tube, the penis, which can be felt under the skin. (Depending on the maturity of your fetal pig) Scrotal sacs can be found in the area ventral to the anus.

2. (Internal Anatomy) Place your pig in the dissecting tray so that the ventral surface is up. Cut two pieces of string about 2 ft. long. Tie a piece of string securely around the ankle of each of the right legs. Run the two pieces of string under the dissecting tray, and tie them tightly to the ankles of the left legs, thus holding the pig securely in place in the dissecting tray, as shown in plate 21.

   A. (ANATOMY) (NOTE: It you stretch the legs as far as possible by tightening the strings during dissection, it will be a lot easier to locate the inner organs)
   Begin your dissection by making an incision through the skin and muscle of the ventral body wall overlying the sternum. Locate the sternum between the forelimbs and make a longitudinal incision as shown in plate 21. Continue the incision anteriorly to the level of the lower jaw and posteriorly to a point just anterior to the umbilical cord. Use a sharp scalpel for your dissection and cut cleanly through the skin and muscles; take care not to damage any underlying organs. Make a second incision from the region of the umbilical cord toward the side of the body, and then connect the first and second incision by cutting around the umbilicus to leave intact a portion of the ventral body wall immediately surrounding the umbilical cord. The fourth incision should be made laterally from the umbilical region toward the side of the pig. Continue your incision in the sequence indicated in plate 21 to provide good access to internal organs of the pig. Remove the skin from the ventral surface of the neck, the lower jaw, and the left side of the head up to the base of the ear. Also, remove the muscle on the ventral side of the larynx and trachea. Use scissors to cut through corners of the mouth to the back of the jaw, so that the mouth opens widely. At completion of these incisions you should be able to observe the following: **head-mouth, hard palate, softpalate, uvula, tongue,** and **teeth; neck-pharynx, larynx, thyroid gland,** and **thymus gland; thorax-lungs, heart,** and **diaphragm; abdomen-liver, gall bladder, stomach, spleen, small intestine, large intestine, kidneys, urinary bladder** and **genital organs.**

3. Draw and label major parts of fetal pig on plate 22.

Questions

1. List and explain two functions of the gall bladder.
2. Why does the intestine have villi?
3. What process allows for movement of nutrients through the alimentary canal?
4. What is the function of the spleen?

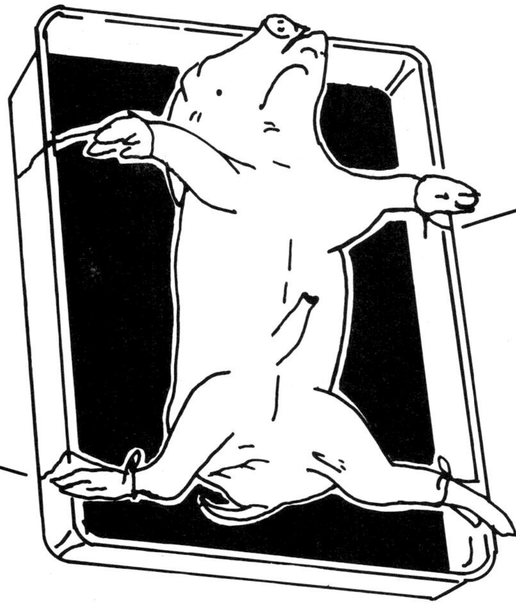
String Connects
Forelegs
String
Connects
Hindlegs

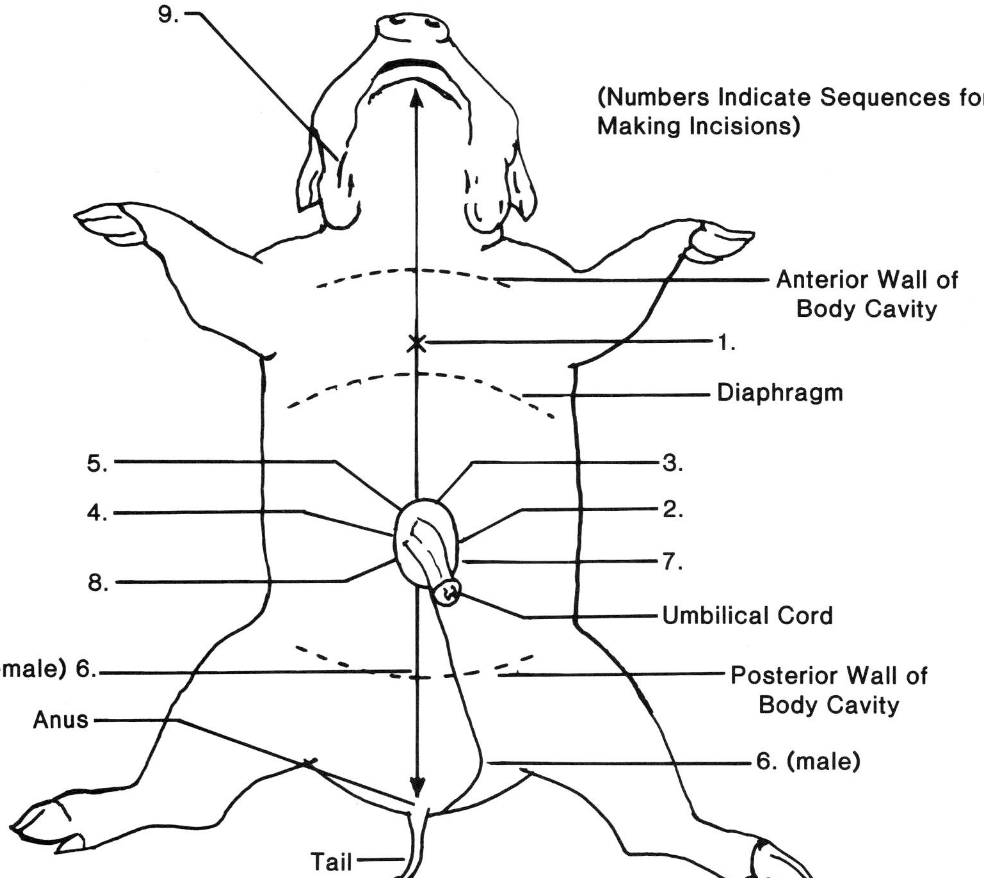
9.
(Numbers Indicate Sequences for
Making Incisions)
Anterior Wall of
Body Cavity
1.
Diaphragm
5.
3.
4.
2.
8.
7.
Umbilical Cord
(female) 6.
Posterior Wall of
Body Cavity
Anus
6. (male)
Tail

Plate 21.

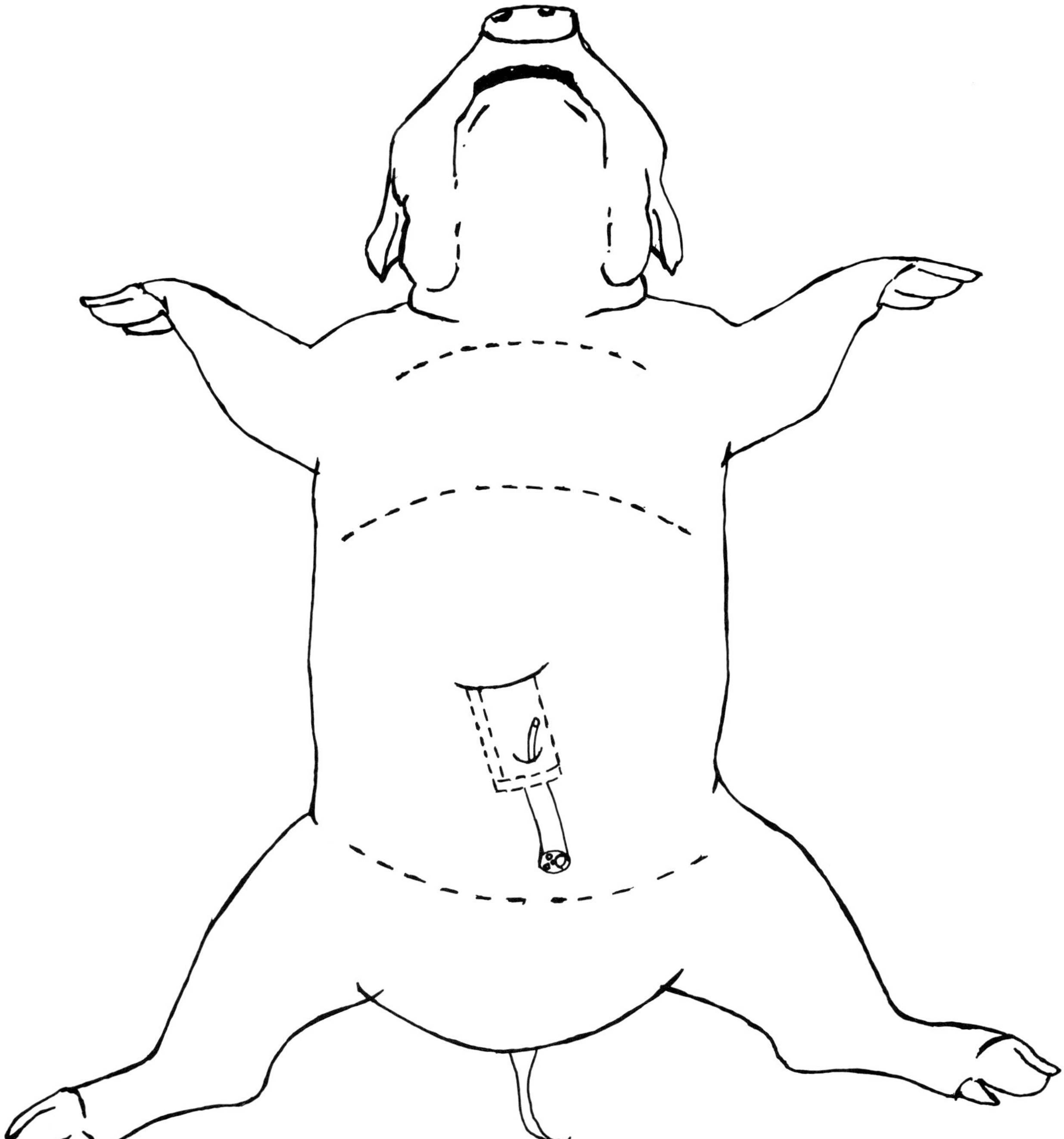

Plate 22.